STUDY GUIDE

to accompany

McCormick * Pressley

EDUCATIONAL PSYCHOLOGY
Learning, Instruction, Assessment

Pamela Beard El-Dinary, Ph.D.
Georgetown University

LONGMAN

An imprint of Addison Wesley Longman, Inc.

New York • Reading, Massachusetts • Menlo Park, California • Harlow, England
Don Mills, Ontario • Sydney • Mexico City • Madrid • Amsterdam

Study Guide
to accompany
McCormick/Pressley, Educational Psychology: Learning, Instruction,
Assessment

Copyright © 1997 Longman Publishers USA, a division of Addison Wesley Longman, Inc.

ISBN: 0-673-99862-2

97 98 99 00 01 9 8 7 6 5 4 3 2 1

CHAPTER 1

An Introduction to Good Thinking and Good Teaching

LEARNING OBJECTIVES

1. Identify characteristics of effective thinking.

2. Describe the direct explanation approach to instruction and give an example.

3. Describe discovery learning and guided discovery and give an example.

4. Describe guided participation and give an example.

5. Describe scaffolding and give an example.

6. Compare and contrast instructional models aimed at developing effective thinking.

7. Explain terms used to describe quantitative and qualitative methods of educational research.

8. Compare and contrast qualitative and quantitative methods of educational research.

9. Describe the grounded theory approach to qualitative research.

10. Evaluate the quality of a qualitative or quantitative research study.

STRENGTHENING WHAT YOU KNOW

The purpose of this chapter is to introduce you to the characteristics of effective thinking and to describe instructional approaches that support effective thinking. The chapter also introduces research methods that are often used to study thinking and teaching. By understanding research methods, you can evaluate the quality of a research study to judge for yourself whether the findings are credible.

Objective 1.   **Identify characteristics of effective thinking.**
Fill in the concept map below and think of an example of each component:

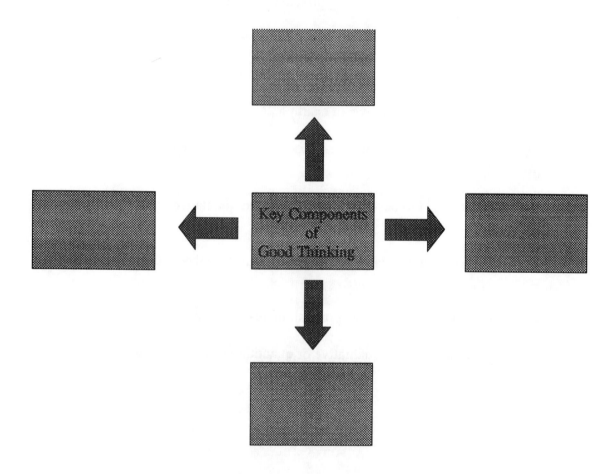

A.   *Strategies*

1. Define strategies and give an example that is meaningful to you.

2. Brainstorm a list of strategies you could use to learn the material for this course.

B.   *Metacognition*

1. Define metacognition and give an example that is meaningful to you.

2. Metacognition has 2 facets--1) knowing about thinking, such as which strategies are effective for which situations; 2) monitoring how well one is thinking, understanding, remembering. Apply these 2 kinds of metacognition right now by a) deciding what kind of strategy can best help you understand and remember these 2 facets of metacognition and b) monitoring how well you understand the meaning of metacognition.

C.   *Knowledge*

1. Give an example of how knowledge can affect one's ability to solve a problem.

2. List some sources of knowledge about learning and teaching that you bring to this course. What kinds of knowledge do you hope to gain from this course?

D. *Motivation*

1. Give an example of how motivation can affect one's ability to solve a problem.

2. How motivated are you to learn the information presented in this course? How is that likely to affect your learning?

E. *Brain Functioning/Short-term Memory*

1. Define short-term memory and give some examples of ways you use short-term memory.

2. Use the concept of short-term memory to explain why you might find it important to purposefully concentrate your attention when studying this material. [Refer to page 9 in the textbook.]

F. *Poor Learners*

1. A student is having difficulties in school. Brainstorm and list at least 10 possible sources of the problem.

Objective 2. **Describe the direct explanation approach to instruction and give an example.**

1. Draw a flow chart showing the sequence of instruction in direct explanation approaches.

2. Explain the 2 most critical features of direct explanation.

3. Develop an example of direct explanation in any setting (classroom, parent/child, professional training...).

Objective 3. **Describe discovery learning and guided discovery and give an example.**

1. Explain the difference between pure discovery learning and guided discovery.

2. Develop an example of pure discovery learning in any setting (classroom, parent/child, professional training...).

3. Modify the above example to make it guided discovery rather than pure discovery.

Objective 4. **Describe guided participation and give an example.**

1. Explain distinctions between guided participation and direct explanation.

2. Develop an example of guided participation in any setting (classroom, parent/child, professional training...).

Objective 5. **Describe scaffolding and give an example.**

1. Explain the relationship between scaffolding and direct explanation, and make a distinction between the two.

2. Develop an example of scaffolding in any setting (classroom, parent/child, professional training...).

Objective 6.   **Compare and contrast instructional models aimed at developing effective thinking.**

1. Complete the following table of Features/Strengths/Weaknesses of the instructional approaches discussed in the text. If you need help, use the samples on the next page; answers may be used more than once. Check your answers against the text.

APPROACH	FEATURES	STRENGTHS	WEAKNESSES
Direct Explanation			
Discovery			
Guided Discovery			
Guided Participation			
Scaffolding			

Sample features [fill in others you can think of]:
Approach emphasizes importance of the right amount of help, not too much or too little.
Approach puts emphasis on explicitly teaching students how to conduct a learning task.
Approach expects students to use steps on their own after being cued to use them.
Approach expects students to internalize effective thinking by interacting with an adult.
Approach takes the emphasis off instruction.
Instruction results from teacher and student continually responding to one another.
Instruction flows from teacher-directed to student-independent.
Amount of explicitness depends on the students' rate of learning.
Tasks give students opportunities to figure out information themselves.
Students practice carrying out the task, with teacher guidance.
Students give rules to define the concepts they are learning.
Students decide for themselves how they will do the task.
Teacher and student work in collaboration to accomplish a task.
Teacher asks questions to lead students to understand ways a problem could be solved.
Teacher limits input to just giving a task and answering student questions about it.
Teacher carefully monitors students' progress and adjusts instruction accordingly.
Teacher gives lots of support at first, giving students more control later.
Teacher gives student subtle hints to guide tasks student can't do alone.
Teacher frequently cues students about what they are supposed to do.
Teacher focuses students' attention to key aspects of a problem.
Teacher explains when, where, and why to use the information.
Teacher gives step-by-step instructions about how to do a task.
Teacher explanations are introduced gradually to students.
Teacher highlights misconceptions students might have.
Teacher explains at the outset ways to carry out a task.
Teacher leads students to important understandings.
Teacher shows examples of effective end products.
Teacher gives the student help when it is needed.
Teacher suggests strategies a student might try.
Teacher models how to do the task.

Sample strengths [add others]
students construct personalized understandings
students internalize metacognitive knowledge
reduces risk of misunderstanding
can cover material in more depth
produces deep understanding
stimulates scientific thinking
promotes student motivation
efficient--learn in less time
can cover more material

Sample weaknesses [add others]
inefficient--more time to learn concept
increases risk that students develop
        incorrect understandings or
        inappropriate strategies
not all students figure out strategies
        they need for the task
requires teacher who can develop
        insightful questions on the spot
leaves more for the child to infer

2. Explain a common feature of all of the instructional approaches in the chapter.
   [See page 12 in your textbook.]

3. Describe how two or more of the instructional approaches can be used together.

4. State an opinion about which instructional approach or combination of approaches you prefer. Describe what the combined approach might look like in a classroom. Defend your opinion with information given about each instructional approach.

Objective 7. **Explain terms used to describe quantitative and qualitative methods of educational research**

1.   *Quantitative methods*
Define and give examples of the following terms:

a) manipulative investigation

k) longitudinal study

b) random assignment

l) cross-cultural study

c) nonmanipulative investigation

m) internal validity

d) mean

n) external validity

e) standard deviation

o) reliability

f) statistical significance

p) confounding variables

g) effect size

q) objectivity

h) factorial design

r) meta-analysis

i) correlation coefficient

j) cross-sectional study

2. *Qualitative methods*
Define and give examples of the following terms:

a) grounded theory (brief definition)

b) member checking

c) theoretical saturation

d) credibility

e) transferability

f) dependability

g) confirmability

3. Explain the difference between manipulative and nonmanipulative investigations.

4. Explain the difference between statistical significance and effect size. Explain why it can be important to know both.

Objective 8.   **Compare and contrast qualitative and quantitative methods of educational research.**

1. Fill in the following chart. Use the terms above to brainstorm points of comparison. Also consider the topics of scientific method/interpretation, the role of subjectivity/objectivity, and the purposes--theory testing/theory constructing.

Qualitative Methods	Quantitative Methods

2. Name several reasons a researcher might chose quantitative methods over qualitative.

3. Name several reasons a researcher might choose qualitative methods over quantitative.

Objective 9.   **Describe the grounded theory approach to qualitative research.**

1. Describe the processes a researcher goes through in developing grounded theory. Think of an example of something you might be interested in developing a theory about (such as a theory of dating, or a theory of the role of recreation in campus life), and tell briefly what you could do at each step to develop a grounded theory.

Objective 10.   Evaluate the quality of a qualitative or quantitative research study.

1.   In the table below, list criteria for evaluating quantitative and qualitative methods (at least 4, preferably more, for each type of research). As much as you can, organize the criteria by matching concerns for quantitative studies to parallel concerns for qualitative studies. Make sure you understand each type of criteria, as well as the distinctions between evaluating quantitative and qualitative studies

Criteria for Evaluating Studies

Qualitative Studies	Quantitative Studies

2. *Triangulation*
   a) Give an example of how triangulation could be used for a quantitative study.

   b) Give an example of how triangulation could be used for a qualitative study.

3. Describe both qualitative and quantitative ways to summarize results across several studies.

PRACTICE TESTS

[See answer key at the end of the chapter for correct responses.]

Multiple Choice

Circle the letter of the best response to each question.

1. Which of the following is **not** a strategy?
    a) knowing that standard deviation indicates differences among scores
    b) forming a mental picture of a researcher in action
    c) taking notes during a lecture
    d) deciding on a research article to critique for a class assignment

2. Which of the following is an example of short-term memory?
    a) remembering the first part of a math problem while reading the rest of the problem
    b) remembering what you ate for breakfast this morning
    c) remembering a definition for a test but forgetting it afterwards
    d) remembering a small piece of information, like a friend's phone number

3. Which of the following features is common to all the instructional approaches the chapter described for developing good thinking?
    a) They all emphasize active learning by giving teachers the role of facilitator who only answers students' questions.
    b) They all involve cuing students about what they are supposed to do.
    c) They all involve teaching students about learning processes as well as content.
    d) They all focus on students constructing their own understandings of content.

4. Which of the following is **not** a feature of direct explanation?
    a) The teacher reduces guidance over time as students become more independent.
    b) Students respond to a question that directs them to developing an explanation of the concept.
    c) Students have opportunities to develop personalized ways of understanding information.
    d) The teacher suggests strategies the students might try to help them with the task.

5. Which of the following features distinguishes guided discovery from pure discovery?
   a) The teacher presents tasks that guide students to figure out the information.
   b) Students decide for themselves how they will do the task.
   c) The teacher highlights misconceptions students might have.
   d) Students are required to infer the concepts themselves.

6. Which of the following is a key feature of guided participation?
   a) The teacher and student work in collaboration to carry out a task.
   b) The teacher frequently reminds students exactly how to carry out the task.
   c) The teacher shows students how to do a task by modeling how he or she does it.
   d) The teacher gives cues that depend on how hard the task is for the student.

7. Which of the following features distinguishes direct explanation from scaffolding?
   a) The teacher gives information from the outset about ways to carry out a task.
   b) The teacher's explicitness depends on the students' rate of learning.
   c) The teacher introduces explanations gradually.
   d) The approach emphasizes importance of giving just the right amount of help.

8. Which of the following approaches takes the emphasis off instruction?
   a) direct explanation
   b) discovery/guided discovery
   c) guided participation
   d) scaffolding

9. Which of the following approaches focuses on when, where, and why to use the information being learned?
   a) direct explanation
   b) discovery/guided discovery
   c) guided participation
   d) scaffolding

10. Which of the following approaches involves sequential instructions for a task?
    a) direct explanation
    b) discovery/guided discovery
    c) guided participation
    d) scaffolding

11. Which of the following approaches focuses on adjusting explicitness of instruction based on the student's current understanding?
    a) direct explanation
    b) discovery/guided discovery
    c) guided participation
    d) scaffolding

12. Which of the following approaches asks students to state rules that explain what they are learning?
    a) guided participation
    b) discovery
    c) scaffolding
    d) guided discovery

13. Which of the following criteria is **not** used to evaluate qualitative research?
    a) transferability
    b) confirmability
    c) credibility
    d) reliability

14. The relationship between human height and weight is an example of which of the following?
    a) negative correlation
    b) standard deviation
    c) positive correlation
    d) mean

15. The relationship between calorie intake and number of pounds lost (assuming all other factors are constant) would be an example of which of the following?
    a) negative correlation
    b) standard deviation
    c) positive correlation
    d) mean

16. A researcher spends a year studying developmental differences of children in India by comparing the strategies used by children ages 5, 7, and 9. Which type of study is the researcher conducting?
    a) cross-cultural
    b) longitudinal
    c) cross-sectional
    d) manipulative

17. Which of the following could potentially weaken a quantitative study?
   a) The researcher did not carefully decide which students would be in which treatment group, simply picking names out of a hat.
   b) The researcher chose a test for which people get different scores in different testing situations.
   c) Rather than looking at the individual scores, the researcher just took the average.
   d) The researcher did not try to make any improvements, simply comparing two groups that already existed.

18. Which of the following could potentially weaken a qualitative study?
   a) Although the people participating in the study disagree with the findings, the researcher decides to stay consistent with the analysis based on previous data.
   b) Instead of being objective, the researcher asks the participants what they think is happening in the situation.
   c) Mid-study, the researcher changes the categories for analyzing data.
   d) The researcher doesn't have any theories at the beginning of the study and therefore starts out without a focus.

19. An instructional study looks at the relationship between amount of teacher prompting and conceptual understanding. The researcher counts teacher prompts and counts students' correct answers on a concept test. The analysis shows a correlation of +.93 between teacher prompts and student conceptual understanding. Which of the following would be an **incorrect** interpretation of these results?
   a) As teacher prompting increases, student conceptual understanding increases.
   b) There is a strong relationship between students' conceptual understanding and teacher prompting.
   c) The way to expand students' conceptual understanding is to provide more teacher prompts.
   d) If you know how much prompting the teacher gave, you can predict how much conceptual understanding the student had.

20. Which of the following would **not** be a confounding variable in study of an instructional approach?
   a) The experimental instruction takes more time than the control instruction.
   b) Students in the control group are shown the materials that are being used with students in the instructional group.
   c) Twenty teachers in a school agree to participate in a study. Ten of them volunteer to teach the experimental intervention, and the other ten serve as a control group.
   d) One group of students gets the experimental instruction, and the control group continues with their normal instruction.

Completion

Fill in each blank with the best fitting term from the chapter. Terms are used only once.

1. Four main components of good thinking are: _____ , _____ , _____ , and _____ .

2. Awareness of one's thinking processes and the ability to monitor one's own performance are included in _____ .

3. A plan of action to aid solving a problem is called a _____ .

4. Statistical analyses are used in _____ research studies.

5. Some students are put in a group getting direct explanation and other students are put in a group getting guided discovery, then the two are compared. This is an example of a _____ quantitative study.

6. An investigation on whether males or females use strategies more often is an example of a _____ quantitative study.

7. Constructing theories based on interpretations of what is happening in a setting, including the viewpoints of the participants, is a focus of _____ studies.

8. _____ theory is constructed through interpretations of data.

9. Researchers asking participants for feedback about whether emerging data analysis categories are plausible is called _____ .

10. When all data are adequately explained, a qualitative study has reached _____ .

Matching A (Instructional Approaches)

Match the letters of the instructional approaches on the right with the numbered descriptions on the left. Some terms may be used more than once.

_____ 1. A parent [P] and child [C] work on a puzzle.
      P: I wonder where this piece could go. What can you do to match the pieces?
      C: I see if the colors match.
      P: Lots of pieces are this color, so it's a problem. There must be another way to match them.
      C: I just use the color, and it's not working. I don't know what to do with this one.
      P: The curve on this piece looks like the opposite of the curve on this piece here. Try it out.
      C: It fits! This other one looks like this here!

_____ 2. A teacher distributes heat-lamps, ice cubes, and a metal tray of worms, telling students to work with the materials then report what they learn.

_____ 3. An executive holds a series of training sessions for new employees to develop an understanding of the company. He presents them with past marketing cases and a set of questions focusing on specific aspects of the company's policy.

_____ 4. A researcher guides her apprentice by having him follow her around during her duties. She starts by telling him the details of what needs to be done for an experiment, thinking out loud about decisions she makes. She then coaches the apprentice to do his own study, often reminding him about decisions he needs to make.

_____ 5. A teacher says, "We've been doing lots of adding, subtracting, multiplying, and dividing numbers. I wonder if it matters which order the numbers go in. How can we find out?

_____ 6. A personal trainer tells his client the sequence of a workout routine. Each session, the trainer calls out the moves and the client does them. In a few weeks, the client does the moves without prompting.

_____ 7. A parent encourages a child to play on the beach, saying, "At the end of the day, I want you to tell me three things you learned about the ocean."

a) direct explanation

b) discovery

c) guided discovery

d) guided participation

e) scaffolding

Matching B (Research Terms)

Match the letters of the description on the right with the corresponding numbered terms on the left. Use each description only once. Some descriptions may be left over.

_____ 1. Confounding variable
_____ 2. Correlation coefficient
_____ 3. Cross-sectional
_____ 4. Effect size
_____ 5. External validity
_____ 6. Internal validity
_____ 7. Longitudinal
_____ 8. Mean
_____ 9. Random assignment
_____ 10. Reliability
_____ 11. Standard deviation
_____ 12. Triangulation

a) Compares people from different age groups at the same point in time.
b) Can tell about the combined impact of variables
c) Tells how much the scores vary among people in a group
d) Controls for variation among participants by giving each person the same chance of being in any treatment group in an experiment
e) Tells how much of a difference there is between two groups' average scores
f) Looks at the same people across a period of time
g) Tells the chances that the researcher would get the same results from one occasion to the next
h) Average of all scores in a group
i) Tells how likely it is that the difference between two groups' average scores would occur by chance
j) Tells whether the results could be interpreted in more ways than what the researcher intended
k) Summarizes relationships between two variables
l) Factor that could influence the results of the study and that is not part of the treatment
m) Could be used to study of the impact of 3 or variables that can be manipulated
n) Tells how much the study is like a real-life situation
o) Uses several approaches to study the same question

LEARNING STRATEGIES

Below are examples of strategies that can help you understand major chapter concepts. Use these examples to guide your own strategies.

Strategy Example #1
Personalize the detective examples of good thinking by (a) imagining a familiar detective character in action, displaying each of the components of good thinking or (b) watching a crime show and identifying components of effective thinking:
- What strategies does s/he use to get evidence, recall crime details, uncover motives?
- How does the detective use metacognition, by either being aware of when a strategy will/won't work, or monitoring how close he or she is to solving the crime?
- What world knowledge does this character have that could help solve the crime?
- What motivates the character to get involved in solving the crime?

Strategy Example #2
Create images that relate the features of instructional approaches to the names.
- For direct instruction, I can imagine the student inside a maze, with the teacher outside telling directions. I imagine that as the student gets closer to the end, the teacher guides rather than just telling the student which way to go.
- For discovery learning, I picture students in lab coats, thinking of them as scientists.
- For guided participation, I think of a staircase, and this helps me remember that this approach cues students to follow certain "steps" for the task.
- For scaffolding, I imagine a building being constructed, with lots of metal frames at the beginning, gradually being removed as the bricks are in place. I think about it being important to have the right amount of scaffolding--too little, and the building cannot stand, too much gets in the way.

Strategy Example #3
Distinguishing between features that affect internal validity and external validity:
First ask, "Does this affect whether the study makes sense **within** itself? Is the study **internally** consistent?" (With a confounding variable you can't even be sure that **in this study,** the results are due only to what the researcher manipulated; too many other things are going on **in** the study.) These are issues of internal validity.
Next ask, "Does this affect whether you could **extend** the results to other situations? Does this seem like an isolated example or one that represents other situations?" Questions of external validity often relate more to the setting, the people involved, and whether the materials and approach could work **outside** a research situation.

23

ANSWER KEY

Strengthening What You Know

Objective/Item
6. 1.
Key to filling in Table

DE Direct Explanation/ **DISC**overy/ **GD** Guided Discovery/ **GP** Guided Participation/ **SCAF**folding

Sample features [fill in others you can think of]
Approach emphasizes importance of the right amount of help, not too much or too little. SCAF
Approach puts emphasis on explicitly teaching students how to conduct a learning task. DE
Approach expects students to use steps on their own after being cued to use them. GP
Approach expects students to internalize effective thinking by interacting with an adult. SCAF
Approach takes the emphasis off instruction. DISC
Instruction results from teacher and student continually responding to one another. SCAF
Instruction flows from teacher-directed to student-independent. DE
Amount of explicitness depends on the students' rate of learning. SCAF (key feature)
Tasks give students opportunities to figure out information themselves. D/GD
Students practice carrying out the task, with teacher guidance. DE SCAF
Students give rules to define the concepts they are learning. GD
Students decide for themselves how they will do the task. DISC, GD
Teacher and student work in collaboration to accomplish a task. SCAF
Teacher asks questions to lead student to understand how a problem could be solved. GD
Teacher limits input to just giving a task and answering student questions about it. DISC
Teacher carefully monitors students' progress and adjusts instruction accordingly. SCAF
Teacher gives lots of support at first, giving students more control later. DE SCAF
Teacher gives student subtle hints to guide tasks student can't do alone. SCAF
Teacher frequently cues students about what they are supposed to do. GP DE
Teacher focuses students' attention to key aspects of a problem. SCAF
Teacher explains when, where, and why to use the information. DE
Teacher gives step-by-step instructions about how to do a task. GP [possibly DE]
Teacher explanations are introduced gradually to students. SCAF
Teacher highlights misconceptions students might have. GD
Teacher explains at the outset ways to carry out a task. DE
Teacher leads students to important understandings. GD SCAF
Teacher shows examples of effective end products. DE
Teacher gives the student help when it is needed. SCAF [often DE]
Teacher suggests strategies a student might try. SCAF DE
Teacher models how to do the task. DE, SCAF

24

6. 1. (cont.)
Sample strengths [add others]
students construct personalized understandings ALL
students internalize metacognitive knowledge DISC
reduces risk of misunderstanding DE, GD
can cover material in more depth DISC/GD
produces deep understanding DISC/GD
stimulates scientific thinking GD
promotes student motivation DISC [others may]
efficient--learn in less time DE
can cover more material DE

Sample weaknesses [add others]
inefficient--takes longer to learn D/GD
increases risk that students will develop incorrect understandings or inappropriate strategies DISC
not all students figure out strategies they need DISC
requires teacher who can develop insightful questions at the right moment GD
leaves more for the child to infer SCAF D/GD

Mulitiple Choice

Correct answers are in bold.
Comments related to other options indicate why that response is incorrect.

1. **a)** This is an example of knowledge.

2. **a)**
    c) is incorrect because short-term memory applies to what you are thinking about at
        a given moment. Remembering something for a test requires a longer span of
        memory--you aren't thinking about the definition every moment, other thoughts
        are pushing it away. Even though you may forget it after what you consider a
        short time, the information is not stored in short-term memory.

3. **d)**
    a) is an example of discovery learning.
    b) is not true of discovery or guided discovery; it represents a key feature of guided
        practice, and cuing is also used as part of scaffolding and direct explanation.
    c) is true of direct explanation.

4. **b)** This is true of inquiry teaching/discovery learning.

5. **c)** is true of guided discovery only; the rest are true of both discovery and guided discovery.

6. **b)** is a feature of guided participation; the rest are features of scaffolding [c) is also a feature of direct explanation.]

7. **a)** true of direct explanation, but not scaffolding; the rest are true of both approaches.

8. **b)**

9. **a)**

10. **c)**

11. **d)**

12. **d)**

13. **d)** Qualitative research uses the term dependability, emphasizing that other people would come to the same conclusions based on the given data set, not that the data would be the same from one time to the next or when gathered by a different person.

14. **c)**

15. **a)**

16. **c)**
    a) Although the study is in India, the researcher is not making comparisons with a different culture.
    b) The study compares different students at each age group, conducted within a year.
    d) The researcher is not giving an intervention, just looking at strategies the students are already using.

17. **b)** represents low reliability
    a) This is an example of random assignment, desirable in quantitative research.
    c) Averages, such as a mean of scores, are typically used to express and compare quantitative findings.
    d) This is an example of a nonmanipulative investigation--a useful approach that is often necessary for studying factors that cannot be controlled, such as age, gender, and race.

18. **a)** Participants' views should be considered in a qualitative analysis, which is why member-checking is often a key step in the research.
    b) Qualitative methods value participants' views. Objectivity is considered impossible to a qualitative researcher; therefore, the goal is to have a rich interpretation based on multiple sources of information.
    c) This is an expected process in developing grounded theory.
    d) Developing theory through data collection is a key feature of qualitative investigations.

19. **c)** Correlation does not imply causation.

20. **b)** This actually helps prevent the materials from being a confounding variable because both groups see the same materials. Then any differences between groups are more likely to be attributed to the type of instruction.
    a) The confound is that simply spending more time with the instructional material could make an impact, in addition to the type of instruction.
    c) In this situation, teacher could be a confounding variable. Instructional differences could depend on the teachers, especially because the kind of teacher who agrees to teach the intervention may already have different teaching approaches or personality qualities than those who don't volunteer. For the sake of the research, ideally all teachers would teach either the experimental instruction or the control instruction. They would be randomly assigned and would not know whether the instruction they are using is the experimental or the control approach.
    d) The confound is novelty; getting something new might have an impact that gives the experimental group an advantage, even if the instruction isn't really better.

Completion

1. knowledge, strategies, metacognition, motivation
2. metacognition
3. strategy
4. quantitative
5. manipulative--also known as an experiment
6. nonmanipulative
7. qualitative
8. grounded
9. member-checking
10. theoretical saturation

Matching A (Instructional Approaches)

__e__ 1.
__b__ 2.
__c__ 3.
__a__ 4.
__c__ 5.
__d__ 6.
__b__ 7.

Matching B (Research Terms)

__l__   1. Confounding variable
__k__   2. Correlation coefficient
__a__   3. Cross-sectional
__e__   4. Effect size
__n__   5. External validity
__j__   6. Internal validity
__f__   7. Longitudinal
__h__   8. Mean
__d__   9. Random assignment
__g__ 10. Reliability
__c__ 11. Standard deviation
__o__ 12. Triangulation

Correct terms for responses left over:
b) factorial design
i) statistical significance
m) factorial design (distractor for triangulation)

28

CHAPTER 2

Motivation

LEARNING OBJECTIVES

1. Explore the role of a learner's self-perceptions and beliefs about learning in his or her motivation to learn.

2. Discuss gender differences in student expectations and attributions.

3. Describe ways to improve learner self-perceptions.

4. Describe ways to facilitate a learner's goal-setting.

5. Explain how learners can use self-instructions to increase their volition.

6. Discuss the role of classroom environment in learners' motivation.

7. Discuss the role of praise and rewards on student motivation; explain how they can be used appropriately, as well as ways that they can undermine motivation.

8. Describe cooperative learning approaches and explain how they can support learner motivation.

9. Discuss the role of cognitive conflict in learner motivation.

10. Discuss the role of interest in motivation; describe ways to develop tasks that maintain attention without distracting the learner from key content.

STRENGTHENING WHAT YOU KNOW

This chapter focuses on factors that affect a learner's motivation, explaining how instruction can make an impact on student motivation. The chapter reveals common situations that can inadvertently decrease learner motivation and explores ways to increase motivation.

Objective 1.   **Explore the role of a learner's self-perceptions and beliefs about learning in his or her motivation to learn.**

    A.   *Self-Efficacy*

What factors influence one's self-efficacy?

What impact can self-efficacy have on learning?

Answer these questions by filling in the diagram below.

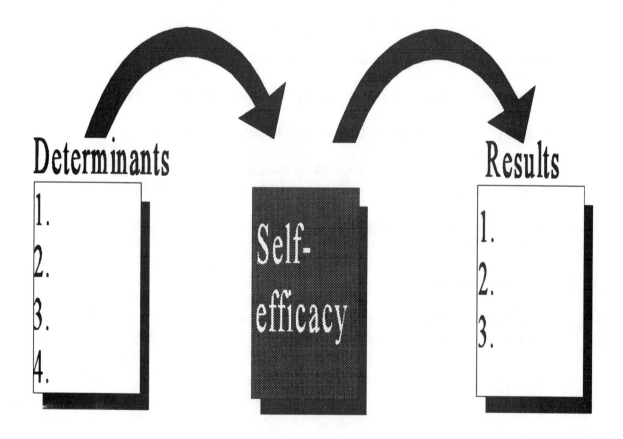

B.   *Possible Selves*

1. What is meant by "possible selves"?

2. What two types of motivation do possible selves provide?

3. List two reasons why is it important to have a realistic possible self.

C.   *Learner Expectations*

1. Draw a line in the following graph to show the pattern of age-related change in student expectations.

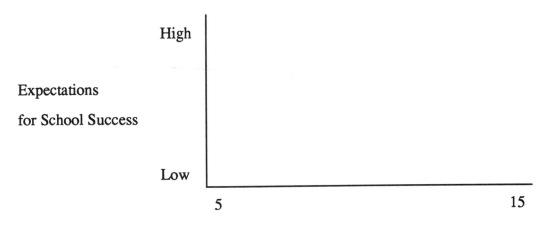

D. *Learner Attributions for Success and Failure*

1. Give your own examples of each of the following kinds of attributions

ATTRIBUTION	Explanation for Failure	Explanation for Success
Effort		
Ability		
Task Factors		
Luck		

2. Which kind of attribution best promotes motivation for future success? Explain why.

3. Define learned helplessness.

4. Explain why learning disabled students are often passive in school.

E. *Learner Perceptions of Intelligence*

1. Describe an entity view of intelligence.

2. Describe an incremental view of intelligence.

Objective 2. **Discuss gender differences in student expectations and attributions.**

1. Explain the differences between boys' and girls' expectations for success, including how expectations change over time.

2. The chapter states that in early elementary school, girls have better math achievement than boys, but later this pattern reverses. List several possible reasons for this.

3. Describe a classroom that supports females' success in mathematics. When might this kind of environment be most critical for a girl? Do you think this type of class would hinder males' math achievement? Defend your position with points made in the chapter.

Objective 3. **Describe ways to improve learner self-perceptions.**
    A. *Increasing Learner Self-Efficacy*
1. What kinds of tasks are best for helping students build self-efficacy? Explain why.

2. Explain when feedback, such as a grade, has the biggest impact on self-efficacy. Contrast this with a situation when a grade may have less impact on self-efficacy.

B.   *Nurturing Possible Selves*

Describe components of the approach for nurturing possible selves discussed in the text.

C.   *Retraining Attributions*

According to the text, what two steps are important in retraining students' attributions?

1.

2.

Objective 4.   Describe ways to facilitate a learner's goal-setting.

A.   *Types of goals*

1. Fill in the following chart with 2 or 3 examples that fit each type.

	General	Specific
Distal		
Proximal		

2. Look at the goals you wrote that are not in the Specific/Proximal box. Try to modify those goals so that they are specific and proximal. For example, break the long-term goals into shorter-term sub-goals, and make the general goals more specific.

B. *Helping Students Set Goals*

Explain the text's four guidelines for helping students set classroom goals, in a way that makes sense to you personally.

1.

2.

3.

4.

Objective 5. **Explain how learners can use self-instructions to increase their volition.**

1. Define volition.

2. Give examples of external and internal distractions that can decrease volition.

3. What are self-instructions?

4. Explain how a learner can use self-instructions to increase volition for a task.

Objective 6.   **Discuss the role of classroom environment in learners' motivation.**

1. Organize chapter information by completing the table below. As you go down each column, you should see how similar concepts relate to one another. As you go across each row, you should see a contrast of "opposites." In each box, include your own explanation of each concept. (The first row of the table has been started for you.)

Reward structures:	personal improvement	competition
Views of intelligence:		
Orientations of classrooms or students:		
Beliefs about being successful:		
Student behaviors related to above views [Fill in based on chapter.]		
Classroom practices that support/relate to the above (may not be equal number in both columns) [Fill in based on chapter.]		

Terms/statements to help fill in table:

Ego involvement	Success depends on effort.	Emphasizing percentile ranks
Incremental theory	Mastery-oriented	Success implies high ability.
Entity theory	Task involvement	Grading on the curve
Making grades public	Discouraged by failure	Failure implies low ability.
More likely to cheat	Grade for improvement	Work hard only if graded
Persistence	Performance-oriented	Compare grades with friends
Discouraged by criticism	Like challenging tasks	Mistakes are part of learning.

2. List several potential problems with competitive classrooms.

Objective 7.   **Discuss the role of praise and rewards on student motivation; explain how they can be used appropriately, as well as ways that they can undermine motivation.**

   A.   *Praise*

In the left column, list features of effective praise.

In the right column, list common mistakes teachers make in giving praise.

Give examples.

Effective Praise	Common Mistakes

B.   *Rewards*

1. In the chart below, write in an explanation of the impact of rewards on intrinsic motivation in the two situations (when intrinsic motivation is initially high, and when it is initially low)

Intrinsic Motivation

	High	Low
Likely Impact of Reward:		

2. Based on the chapter, what guidelines would you give a teacher who is considering whether to use a reward in a specific situation?

Objective 8.   Describe cooperative learning approaches and explain how they can support learner motivation.

A.   *Social structures in classrooms*

Give a brief definition and example of the following social structures in classrooms:

Competitive	Individualistic	Cooperative

B. *Cooperative learning*

1. List the four essential characteristics for cooperative learning, identified by Johnson and Johnson.

a)

b)

c)

d)

2. Research by Fantuzzo, King, and Heller (1992) and by Slavin (1985) showed that the combination of group rewards and individual accountability makes cooperative learning most effective. Why do you think this is true?

Objective 9.   **Discuss the role of cognitive conflict in learner motivation.**

1. Define cognitive conflict.

2. Explain how cognitive conflict can encourage learning of a new concept.

3. How can teachers use cognitive conflict in their instruction?

Objective 10. **Discuss the role of interest in motivation; describe ways to develop tasks that maintain attention without distracting the learner from key content.**

A. *The paradox of interest*

Describe the paradox of the role of interest in learning.

B. *What makes a text interesting?*

1. Describe several features that make a text interesting.

2. What can happen when a fascinating detail is inserted into a textbook?

C. *Computer games and motivation*

1. List four characteristics that make educational computer games motivating.

a)

b)

c)

d)

2. What advice would you give to help teacher who is selecting educational software?

PRACTICE TESTS

[See answer key at the end of the chapter for correct responses.]

Multiple Choice

Circle the letter of the best response to each question.

1. Which of the following attributions for success and failure is most likely to foster success?
   a) task difficulty
   b) effort
   c) ability
   d) luck

2. Which of the following attributions is personally controllable?
   a) task difficulty
   b) effort
   c) ability
   d) luck

3. Which of the following is the most likely action for a student who has developed learned helplessness in sports?
   a) I talked to my school counselor to try to get in a class that has a supportive physical education teacher.
   b) The week before I was going to be graded in basketball, I practiced every night after school because I know I'm not very good.
   c) When I got a bad grade in basketball, I told myself it was because I was up against a tough team.
   d) When it was my turn to shoot baskets for a grade, I just threw the ball in the air without aiming.

4. Which of the following teacher responses (said to the student in private) is most likely to support motivation in the student who is learned helpless?
   a) (The student makes a basket.) "Oh my gosh! You really did it!"
   b) (The student misses.) "That's okay! I know you gave it your best shot."
   c) (The student makes a basket.) "Hey! You're really great at this!"
   d) (The student misses.) "I know you can do this if you just practice more and concentrate on the shot."

41

5. Which of the following is the least likely explanation for girls having lower expectations than boys for success in math?
   a) Girls are not as good in math as boys are.
   b) Girls are more easily frustrated by failure than are boys.
   c) Parents expect higher math grades of boys than of girls.
   d) Math classes may include lots of public drill and practice.

6. If all other factors were the same, which of the following students would probably have the lowest expectations for school performance?
   a) A second grade boy
   b) A second grade girl
   c) A fifth grade boy
   d) A fifth grade girl

7. Which of the following classroom features would **not** be likely to support girls' achievement?
   a) Telling students that math is important for both boys and girls
   b) Arranging teacher-student conferences
   c) Displaying a list of the top 10 students
   d) Scheduling an all-girl class

8. Which of the following students would benefit most from a reward?
   a) Salamon is not interested in doing chemistry experiments, and he has repeatedly found excuses to miss class on lab days. The teacher decides to drop one of Salamon's favorite jelly beans in a test tube for each lab Salamon completes; when the test tube is full, he gets to take the jelly beans.
   b) Casey has been making up songs that help her remember science concepts taught in class, and she has shared them with the teacher to tell the class. The songs are so clever that the teacher tells Casey she will get extra credit for each future song she writes.
   c) Ricke refuses to participate in singing during music class. One day after class, the teacher says to Ricke privately, "I really need your participation in class. If you will participate, I'll give you a homework pass any week you sing every day."
   d) Mr. Hubbard wants to develop a student magazine in which students publish their writing. He is new to the school, and he does not know if the students like to write. To make sure he will get enough responses, he advertises a pizza party for all students whose writing is selected.

9. Which of the following is **not true** about rewards?
   a) Promising a reward for activities students like helps keep their motivation high.
   b) Giving a student a reward for a behavior can increase the chances that the student will repeat the behavior.
   c) If a reward seems like a bribe, motivation is likely to decrease.
   d) If a student is highly reluctant to do the task, a reward can give the student enough motivation to get started.

10. Of the following possible selves, which would probably be the best academic motivator for an average-achieving student?
   a) Professional stunt person
   b) Fast food cashier
   c) President of the United States
   d) Research engineer

11. Which of the following was **not** true of the training package designed to nurture students' possible selves?
   a) The training included many easy jobs that would be realistically attainable for the student population.
   b) The training included an emphasis on attributions, by building students' understanding that they controlled their success through academic efforts.
   c) The training focused on how to cope with failure.
   d) The training was aimed at increasing awareness of a variety of vocations.

12. Students get points for each correct answer on each test. For one test, the teacher gives each student a piece of candy for each point the student earns. What type of social situation does this example describe?
   a) cooperative
   b) competitive
   c) individualistic
   d) conflictive

13. Students get points for each correct answer on each test. The teacher promises a pizza party whenever the class accumulates 10,000 points. What type of social situation does this example describe?
   a) cooperative
   b) competitive
   c) individualistic
   d) conflictive

14. Students get points for each correct answer on each test. The teacher gives a homework pass to the student with the top score. What type of social situation does this example describe?
   a) cooperative
   b) competitive
   c) individualistic
   d) conflictive

15. Which of the following is **not** an essential characteristic of cooperative learning, as identified by Johnson and Johnson?
   a) The task must require everyone to help.
   b) Each person has to be accountable for doing the task required.
   c) The group members must not argue, but should agree on the answers.
   d) The teacher has to teach the class how to behave in a small group.

16. Which of the following is true of cognitive conflict?
   a) It confuses students so that they become unable to learn a new concept.
   b) It motivates students to try to understand a new perspective.
   c) It has little impact on a person's existing knowledge.
   d) Teachers should avoid this kind of conflict.

17. Which of the following is a correct instructional implication of cognitive conflict?
   a) There is a specific technique teachers follow in instruction based on cognitive conflict.
   b) Teachers should encourage students to express their inaccurate beliefs.
   c) In cognitive conflict approaches, the teacher lets students discover the correct concept for themselves.
   d) Teachers should just call on students who understand the concept, so they can explain it correctly to the rest of the class.

18. Which of the following is a feature of effective praise?
   a) Praise emphasizes the student's class participation.
   b) Praise is given before the teacher wants the student to do something.
   c) Praise is given even if the student didn't really earn it.
   d) Praise includes a suggestion that the student tried hard.

19. Of the following examples of teacher praise, pick the most effective.
   a) "I really appreciate your contribution. Thanks for sharing that with the class."
   b) "This is the very best essay I have ever read in my entire life. I am so impressed with your work."
   c) "I like how your answer contrasted the two poems. You studied the new poem and thought about how it fit what we were learning."
   d) "You have so much potential in this class. If you review your class notes each night, I have confidence that you will excel on the test next week."

20. Pick the true statement about interest.
   a) An interesting detail in a textbook can increase learning of main ideas.
   b) Among educational software choices, computer games are better than drill-and-practice.
   c) Interest can both support learning and distract from learning.
   d) Students prefer to read about new experiences and people who are different from themselves.

Completion

Fill in each blank with the best fitting term from the chapter. Terms are used only once.

1. A student who believes that intelligence is genetic is a(n) _____ theorist.

2. A student who believes that environment affects intelligence is a(n) _____ theorist.

3. Cooperative learning is most effective when it includes _____ rewards and _____ accountability.

4. Promising an external reward when the student already enjoys the activity can produce the _____ effect.

5. If one student succeeds, others fail in _____ social contexts.

6. One person's success has no impact on whether other students succeed or fail in _____ social contexts.

7. An individual's success depends on the success of others in _____ social contexts.

8. Writing because you enjoy creating and communicating ideas is an example of _____ motivation.

9. Reading because you get to go to a party when you read ten books is an example of _____ motivation.

10. _____ details can draw attention away from key points.

Matching

Match the letters of the description on the right with the corresponding numbered terms on the left. Use each description only once. Some descriptions may be left over.

_____ 1. Attribution
_____ 2. Cognitive conflict
_____ 3. Ego involvement
_____ 4. Intrinsic motivation
_____ 5. Learned helplessness
_____ 6. Overjustification
_____ 7. Possible self
_____ 8. Self-efficacy
_____ 9. Task involvement
_____ 10. Volition

a) Discrepancy between current knowledge and new knowledge.
b) Fantasy about becoming something extraordinary
c) Possible impact of a reward when internal motivation is high
d) Focus on feeling successful when performing better than others
e) Personal desire to do an activity
f) Action that increases the chance that the behavior preceding it will happen again.
g) One's attainable dream or aspiration
h) "There's nothing I can do to understand this."
i) Focus on feeling successful when improving one's own performance
j) How a student explains performance on a task
k) Ability to persevere through a task
l) Belief that one has what it takes to reach a goal

LEARNING STRATEGIES

Below are examples of strategies that can help you understand major chapter concepts. Use these examples to guide your own strategies.

Strategy Example #1
Use imagery to distinguish between intrinsic and extrinsic motivation. Think of a time you did something because it was fun and enjoyable for you; associate this image with a big "I" for intrinsic. (Think of the "I" as standing for "**I** like to do this.") To remember extrinsic motivation, think of a time you did something because you were getting a reward, like points or a prize; associate this image with a big "X" for "extrinsic." The "X" can also symbolize that extrinsic motivators can be detrimental when intrinsic motivation is already high. (You might think of the X "crossing out" or canceling out the I.)

Strategy Example #2
To remember the concept of learned helplessness, think about the way circus elephants are trained so that they don't run away. The baby elephants have heavy chains on their feet so that they can't go anywhere but where the trainer allows. At first, the baby elephants may try hard to run, but their efforts keep failing. After awhile, the elephants don't try to run away anymore. The trainer uses lighter chains, but the elephants still don't try, so they don't know they can escape. Eventually, there are no chains at all, but the elephants have learned that they are helpless, and they don't try at all. This is parallel with what happens with students and the processes through which they become learned helpless in academics through repeated failure.

Strategy Example #3
To remember the concepts of entity theory and incremental theory of intelligence: Think of the meaning of the word "entity" as a thing or object. Imagine a bunch of cartoon people, standing in a line, holding out their hands. As they go through the line, somebody hands each of them a brain, and the characters go off carrying around the brains they were given. This is an image of entity theorists, who believe that intelligence is a thing you are born with that can't be changed.

Think of "incremental" as a series of steps. Imagine a bunch of people, each climbing up a staircase. You might want to imagine their brains getting larger as they climb each step. This can help remind you that incremental theorists believe they can take action to improve their intelligence, and that intelligence builds up progressively through experience.

ANSWER KEY

Strengthening What You Know

Objective/Item
1. A.   Determinants--previous success, social models, other's opinions, feedback.
         Results--effort, goal-selection, persistence.

1. B. 2. Direction and Energy

1. B. 3. An unrealistic possible self can lead motivation in an unproductive direction.
          A realistic possible self can be powerful motivator.

3. A. 2. Feedback has most impact when the learner is attempting new goal.
          If the learner already has strong self-efficacy, feedback is less likely to have an
          impact.

Multiple Choice

Correct answers are in bold.
Comments related to other options indicate why that response is incorrect.

1. **b)**

2. **b)**

3. **d)**
    a) This shows interest in succeeding.
    b) effort
    c) task difficulty

4. **d)**
    c) Encourages ability attribution which isn't the ideal anyway; most likely the
        student will not "buy it."

5. **a)**
    d) See page 33 in textbook.

6. **d)**

7. **c)** [Competition]

8. **a)**
    b) overjustification effect
    c) bribe
    d) The teacher does not know if students are intrinsically motivated to write. It is better to see if students are interested in writing before using a reward.

9. **a)** If the reward is suggested before performance, there is a risk of replacing students' intrinsic enjoyment with their desire to get a reward. If the reward is suggested only after performance, it can be an added bonus that supplements the students' personal motivation for the task.

10. **d)** more realistic goal, but still gives student a way to go
    a) low probability & takes motivation away from academics
    b) probably not a high enough goal; student probably doesn't need much progress to attain it
    c) low probability, although a worthy goal

11. a) The point of possible selves is not to have an easy alternative, but something that could be realistically attained with lots of effort.

12. **c)**
    d) [distractor--no such term in text]

13. **a)**
    d) [distractor]

14. **b)**
    d) [distractor]

15. **c)** One way students learn in cooperative groups is by the cognitive conflict that comes about when students are at different levels of understanding.

16. **b)**

17. **b)**

18. **d)**
   a) Should emphasize the participation that supports learning, not just participation.
   b) Praise is given after the student does something the teacher wants.

19. **c)**
   a) focuses on participation, but not on how it supports learning.
   b) sounds insincere; not specific about what was good about the essay
   d) not contingent on behavior; instead this praise mostly anticipates desired behavior. It might be motivating, but praise for something the student has accomplished may be more meaningful.

20. **c)**
   a) can actually decrease, calling attention away from key points
   b) games are more motivating, but not always more educational

Completion

1. entity
2. incremental
3. group rewards/individual accountability
4. overjustification
5. competitive
6. individualistic
7. cooperative
8. intrinsic
9. extrinsic
10. seductive

Matching

  __j__   1.   Attribution

  __a__   2.   Cognitive conflict

  __d__   3.   Ego involvement

  __e__   4.   Intrinsic motivation

  __h__   5.   Learned helplessness

  __c__   6.   Overjustification

  __g__   7.   Possible self

  __l__   8.   Self-efficacy

  __i__   9.   Task involvement

  __k__ 10.   Volition

Correct terms for responses left over:

b) none--distractor for possible selves

f) reward/reinforcer

CHAPTER 3

Representation of Knowledge

LEARNING OBJECTIVES

1. Explain concepts as representations of knowledge, including how concepts are formed and how they are interrelated in memory.

2. Describe the educational implications of the model of semantic networks of concepts.

3. Explain propositions as representations of knowledge, including how propositions are formed and how they are interrelated in memory.

4. Describe the educational implications of propositional networks.

5. Explain the connectionist model of neural networks.

6. Describe the educational implications of the neural network model.

7. Explain schema theory, including schema activation.

8. Describe the educational implications of schema theory, including how people use schemata to understand information.

9. Explain the dual coding model and its educational implications.

10. Explain productions as representations of procedural knowledge, including how procedures are learned and the educational implications of this kind of learning.

11. Compare and contrast the models of knowledge representation.

12. Explain the difference between episodic and semantic memory.

STRENGTHENING WHAT YOU KNOW

The purpose of this chapter is to explain different models of how knowledge is formed, represented, and organized in long-term memory. The chapter applies theories of knowledge representation to explain how students learn and understand information and tasks.

Objective 1.   **Explain concepts as representations of knowledge, including how concepts are formed and how they are interrelated in memory.**

    A.   *Role of concepts*

1. Define the term "concept."

2. How do concepts help us organize our experiences? Give a specific example.

    B.   *Concept formation*

1. Describe the feature comparison theory of concept formation. Give an example.

2. Describe the prototype theory of concept formation, including how prototypes are formed. Give an example.

3. What is a "fuzzy concept"? Which of the above theories is most useful for categorizing fuzzy concepts? Why?

    C.   *Relationships among concepts*

1. "Define semantic networks." What are the "nodes" and "links" in a semantic network?

2. What is meant by spreading activation?

Objective 2.   **Describe the educational implications of the model of semantic networks of concepts.**

   A.   *Activating Background Knowledge*

1. Using the terms about semantic networks and concepts, explain why it is helpful to encourage students to activate their background knowledge.

2. Now use your own words to explain how this helps.

3. Based on the hierarchical organization of concepts, explain why some ideas about a topic might be harder to remember than others.

4. Give a specific example of how a teacher could directly teach a concept. Use the theories of concept formation to explain how this works.

5. Create a specific example of how a teacher could use discovery learning to teach a concept. Use the theories of concept formation to explain how this works. Compare this explanation to the one you created for directly teaching a concept.

Objective 3.   **Explain propositions as representations of knowledge, including how they are interrelated in memory.**

    A.   *Propositions*

1. Define proposition. Give some examples.

2. Define microproposition

3. Define macroproposition

4. Define "bridging inference."

Objective 4.   **Describe the educational implications of propositional networks.**

    A.   *Text comprehension and memory*

1. Explain the processes of understanding a text, according to the model of propositions.

2. According to the model of propositions, what kinds of information is a person likely to mention when asked to recall a class lecture? What would probably be remembered first? What else might be included?

3. A few weeks after a class discussion, you suddenly think about an opinion one of your classmates stated. How might this memory be explained by propositions?

Objective 5.   **Explain the connectionist model of neural networks.**
   A.   *Neural networks*
Describe the three basic components of neural networks.
1.

2.

3.

Objective 6.   **Describe the educational implications of the neural network model.**
1. Explain how learning occurs according to the neural network model.

2. Give an explanation, in terms of the model, for why lower-order knowledge must be mastered before higher-order learning can take place.

3. Define two types of practice and explain why one is better than the other.

Objective 7.   **Explain schema theory, including schema activation.**

1. Define schemata.

2. What are variables or slots in a schema?

3. What does "instantiated" mean in schema theory?

Objective 8.   **Describe the educational implications of schema theory, including how people use schemata to understand information.**

1. Explain how activating a schema could affect one's understanding of a task. Include the role of schemata in making inferences, focusing attention, and recalling information.

2. If students activate their schemata to help them remember something, how might this backfire?

3. Define narrative text structure and its components.

4. Define expository text structures and give some examples.

Objective 9. **Explain the dual coding model and its educational implications.**

    A.   *Images*

1. List four kinds of images included in the dual coding model.

    a)

    b)

    c)

    d)

2. What is the role of images in the dual coding model?

3. How could you develop instruction to capitalize on dual coding? Give one or two examples.

Objective 10. **Explain productions as representations of procedural knowledge, including how procedures are learned and the educational implications of this kind of learning.**

   A.   *Procedural versus declarative knowledge*

Fill in the following chart to contrast declarative and procedural knowledge:

	Procedural Knowledge	Declarative Knowledge
Definition:		
Examples from my life:		
How can this type of knowledge be demonstrated?		
Can the truth of this type of knowledge be determined?		
How much practice is needed to acquire this kind of knowledge?		

   B.   *Productions*

1. Define "production." Give an example.

2. Explain pattern recognition productions and why they are important. Give an example.

C. *ACT**

1. What is ACT*?

2. List 3 ways declarative knowledge is encoded according to ACT*
    a)

    b)

    c)

3. Explain how procedural knowledge is acquired according to ACT*.

D. Instructional implications

1. Why does learning how to do a task start out slowly, later becoming quicker and easier?

2. Explain two problems students can have in learning when their are discrepancies between declarative and procedural knowledge.

3. What is meant by situated knowledge? What does it imply for education?

61

Objective 11.   Compare and contrast the models of knowledge representation.

1. Complete the following table. Fill in as much as you can on your own, then refer to the items on the next page. Some items will appear in more that one row of the table.

Type of knowledge	Model	Basic Building Block of Knowledge	Organization	Implications for Learning
(A)	Concept formation			
(B)	Propositions			
(C)	Connectionist			
(D)	Schema Theory			
(E)	Dual Coding Theory			
(F)	Productions			

Types of knowledge: Declarative   Procedural

Building blocks of knowledge:
productions   concepts   schemata   verbal representations   propositions   images
smallest unit of meaning that can be judged as true or false
connections between units of information, rather than the units themselves
larger chunks of knowledge than other models
mental representation of a category of related items
generalized knowledge about objects, situations, or events

Organization:
Semantic network   Neural network   Propositional network
Hierarchical   Ideas linked together in sequence
Describes connections between units of information
Nodes are concepts, with links showing relationships.
Perceptual information linked to related verbal information
Nodes are units of information that can be activated at a conscious or subconscious level.
Slots are filled in by specific instances of the current information.
Spreading activation
Links are through common elements.
Knowledge grouped as a skeleton of information about the topic
An information unit is activated only when linked information is activated.

Instructional implications:
Learning is essentially through building and strengthening patterns of connections.
Thinking about what you already know about a topic can assist learning of related ideas.
When remembering, people are more likely to connect ideas closely related in a category.
Learners put details together to get the gist of text; prior knowledge helps understanding.
Learning a new task will be slow at first, but with time the task will be quicker & easier.
Concrete materials are more memorable than abstract materials.
Learners make bridging inferences that are not stated in text, and when asked to recall the
        text, they may remember the bridging inferences as well as main ideas in the text.
People remember typical category members more easily than atypical category members.
Spreading activation occurs among basic idea units that are linked by common elements.
Practice is important to learning.
It is critical to help students make connections among ideas.
Categorizing new information may be based on comparing with a typical example
        (prototype), or by checking necessary and sufficient features.
Learners must master lower order knowledge before higher order knowledge.
Previous understandings of a topic determine how new information is processed.
It may be useful to employ multiples senses when learning new information.

2. Define advance organizers. Choose any of the first 4 theories in the chart (A-D), and use it to explain why advance organizers work.

3. Explain the effectiveness of self-questioning based on any of the first 4 theories in the chart (A-D).

Objective 12.   **Explain the difference between episodic and semantic memory.**
1. Fill in the following table to distinguish between episodic and semantic memory.

	Semantic Memory	Episodic Memory
Point of reference?		
Coded with respect to when it was acquired?		
How easily forgotten?		
"Known" or "Remembered"?		
Big concern of educators?		

PRACTICE TESTS

[See answer key at the end of the chapter for correct responses.]

Multiple Choice

Circle the letter of the best response to each question.

1. Which of the following terms best describes the organization of concepts?
   a) neural
   b) activational
   c) sequential
   d) hierarchical

2. Knowledge of familiar events may be best represented as:
   a) schemata.
   b) semantic memory.
   c) concepts.
   d) productions.

3. Which of the following theories best explain why people often include non-defining features when classifying concepts?
   a) macropropositions
   b) feature comparison
   c) episodic
   d) prototype

4. A mental connection between pictures and words is a simplified description of which of the following theories?
   a) imaging
   b) dual coding
   c) semantic memory
   d) situated knowledge

5. Connections among units of information are best represented in:
   a) semantic networks.
   b) neural networks.
   c) propositional networks.
   d) nodes.

6. Knowledge about changing the oil in a vehicle is probably best represented in
   a) images.
   b) procedures.
   c) schemata.
   d) productions

7. Studies of mental rotation support which of the following models?
   a) dual coding
   b) productions
   c) schemata
   d) procedural networks

8. Which of the following explains how connections between procedures are strengthened?
   a) You repeatedly talk yourself through a task, saying the steps out loud.
   b) You try to remember the last time you did the task, and think about what you did as you work on it this time.
   c) You go through a set of steps to do a task, and it works.
   d) You try to get a visual image of doing the task correctly, and at the same time you say the steps out loud.

9. Which of the following characterizes semantic memory?
   a) The self is the main point of reference.
   b) It is organized knowledge about the world.
   c) The time when the information was acquired is important.
   d) Of the two types of memory, it is most likely to be forgotten.

10. The distinctions a person makes between "short-order cook" and "chef" can be best explained by which view of how concepts develop?
   a) prototype
   b) defining features
   c) episodic
   d) semantic

11. Which of the following characterizes episodic memory?
   a) It is important to ask, "Is this true about the world?"
   b) Teachers are usually most concerned with this type of memory.
   c) It is more appropriate to think of it as remembered than as known.
   d) It is based mostly on major, memorable events.

12. Classifying a person as a teacher because he or she works in a school, develops instruction, and presents lessons to children is an example of which perspective of concept formation?
    a) prototype
    b) defining features
    c) episodic
    d) semantic

13. The ACT* knowledge system says <u>declarative knowledge</u> is represented in all of the following, <u>except</u>:
    a) Images.
    b) Time sequence.
    c) Propositions.
    d) Schemata.

14. Which of the following is **not** true about propositions?
    a) Each one has a subject (or object) and a verb (or preposition).
    b) They can be judged as true or false.
    c) They are sentences in one's mind that describe relationships.
    d) It may take more than one proposition to express an idea.

15. Which of the following come directly from a text?
    a) micropropositions $
    b) schemata
    c) episodes
    d) macropropositions

16. According to the model of propositions, which of the following pairs of ideas are likely to be remembered together?
    a) Cognitive conflict can motivate students to understand new information.
       It can be helpful to let students disagree about a concept.
    b) Effort attributions promote motivation.
       Learned helpless students give up on school.
    c) Imagining yourself in an exciting career can motivate you to complete your schoolwork.
       Metacognition includes monitoring one's own learning processes.
    d) A reward can backfire if a student has intrinsic motivation for a task.
       Intrinsic motivation is the student's own interest in doing a task.

17. Which of the following is **not** a basic component of a neural network model?
    a) neuron
    b) Creating and strengthening of connections define learning.
    c) Connections either excite or inhibit two units simultaneously.
    d) node

18. Elementary students often find it funny to give "complete" addresses, like--
    Jamie's room, 202 N. Street, Centerville, IN, USA, Earth, The Solar System,
    Milky Way, ... What organizational view might best explain this kind of
    thinking?
    a) schematic network
    b) propositional network
    c) neural network
    d) semantic network

19. Which of the following does the model of situated knowledge suggest?
    a) Explicit knowledge can be directly presented to students.
    b) Learning occurs in real-life situations.
    c) After ten years in a profession, a person basically knows all there is to learn.
    d) Knowledge is separable from the learning environment.

20. Which of the following is **not** true of schemata?
    a) Involves putting details together to develop the main idea
    b) Can lead to "remembering" information that was not actually presented
    c) Are like skeleton structures of objects, events, or situations
    d) Often determined by events that recur in life

Completion

Fill in each blank with the best fitting term from the chapter. Terms are used only once.

1. In neural networks, learning occurs when connections are created between
   _____.

2. In situation models, _____ are integrated with the person's background knowledge.

3. Children are less likely than adults to rely on _____ concept knowledge.

4. A model of thinking that includes both declarative and procedural knowledge is called _____.

5. Young children and children without formal schooling often prefer to categorize ideas according to _____ relations.

6. Macropropositions are derived from _____.

7. Procedures develop from declarative representations through _____.

8. When making mental connections among concepts, _____ moves first to concepts more highly associated in the hierarchy.

9. One effective way to build connections in one's knowledge may be _____ practice.

10. The _____ model of knowledge representation suggests that lower-order knowledge must be mastered before higher-order knowledge can be learned.

Matching

Match the letters of the description on the right with the corresponding numbered terms on the left. Use each description only once. Some descriptions may be left over.

_____ 1. Images
_____ 2. Macropropositions
_____ 3. Neural networks
_____ 4. Nodes
_____ 5. Productions
_____ 6. Propositions
_____ 7. Propositional networks
_____ 8. Schemata
_____ 9. Semantic
_____ 10. Semantic networks

a) Non-verbal representations that resemble perceptual experience
b) Descriptions that specify relationships between things and properties of things
c) Gist
d) How concepts are organized
e) Represent familiar situations and determine how information is processed in such situations
f) Organization of knowledge units that can be judged as true or false
g) Connections among nodes
h) If-then
i) Based on a typical member of a category
j) Memory of personally experienced events
k) Meaning units derived directly from a text
l) Memory that includes organized knowledge about the world
m) Connected by simultaneous excitation or inhibition

LEARNING STRATEGIES

Below are examples of strategies that can help you understand major chapter concepts. Use these examples to guide your own strategies.

Strategy Example #1
Draw diagrams to compare semantic networks, propositional networks, neural networks, dual coding, and productions. Include an explanation and example for each part of the diagrams.

Strategy Example #2
Select a content area (the one you plan to teach, if you have one). Identify types of schemata students my be able to apply to this content area? What schemata will students already have? What schemata would you need to help them develop?

Strategy Example #3
How have your schemata about school and teaching affected your understanding of concepts in this course so far?

ANSWER KEY

Strengthening What You Know

Objective/Item

4. A. 3.   Maybe you hear an idea that shares a common element with what your classmate said. For example, someone else uses a term that the classmate used in his or her argument.   Another explanation is that you come across something related to a main idea of that lesson (part of the macrostructure), and this thought activates the thought about the classmate's comment (which was also a main idea, part of the macrostructure).

11. 1.

Type of knowledge	Model	Building Block of Knowledge	Organization	Implications for Learning
(A) Declarative	Concept formation	concepts mental representation of a category of related items	Semantic network Hierarchical Nodes are concepts, with links showing relationships. Spreading activation Describes connections between units of information	Categorizing new information may be based on comparing with a typical example (prototype), or by checking necessary and sufficient features. People remember typical category members more easily than atypical category members. When remembering, people are more likely to connect ideas closely related in a category. It is critical to help students make connections among ideas. Thinking about what you already know about a topic can assist learning of related ideas.

(B) Declarative	Propositions	propositions smallest unit of meaning that can be judged as true or false	Propositional network Links are through common elements. Spreading activation Describes connections between units of information	Learners put details together to get the gist of text; prior knowledge helps understanding. Learners make bridging inferences that are not stated in text; when asked to recall the text, they may remember these inferences as well as main ideas in the text. Spreading activation occurs among basic idea units that are linked by common elements. It is critical to help students make connections among ideas. Thinking about what you already know about a topic can assist learning of related ideas.
(C) Declarative	Connectionist	connections between units of information, rather than the units themselves	Neural network Nodes are units of information that can be activated at a conscious or subconscious level; activated only when linked information is activated. Describes connections between units of information	Learning is essentially through building and strengthening patterns of connections. Learners must master lower order knowledge before higher order knowledge. It is critical to help students make connections among ideas. Thinking about what you already know about a topic can assist learning of related ideas. Practice is important to learning.
(D) Declarative	Schema Theory	schemata generalized knowledge about objects, situations, or events; larger chunks of knowledge than other models	Knowledge is grouped as a skeleton of information about the topic. Slots are filled in by specific instances of the current information.	Previous understandings of a topic determine how new information is processed. Thinking about what you already know about a topic can assist learning of related ideas. It is critical to help students make connections among ideas.

(E) Declarative	Dual Coding Theory	images verbal representations	Perceptual information linked to related verbal information	Concrete materials are more memorable than abstract materials. It may be useful to employ multiples senses when learning new information. It is useful to help students make connections between verbal and visual representations of ideas.
(F) Procedural	Productions	productions	Ideas linked together in sequence	Practice is important to learning. Learning a new task will be slow at first, but with time the task will be quicker and easier.

Multiple Choice

Correct answers are in bold.
Comments related to other options indicate why that response is incorrect.

1. **d)**

2. **a)**

3. **d)**

4. **b)**

5. **b)**

6. **d)**

7. **a)**

8. **c)**
   a) This could explain how procedures can develop from declarative knowledge, but it does not focus on making connections among procedures.

9. **b)**

10. **a)**

11. c)
    d) No, includes all life events and is generally easily forgotten.

12. **b)**

13. **d)**

14. **c)** They do describe relationships, but mentally they are ideas rather than sentences.

15. **a)**

16. **d)** According to the propositional model, these ideas would be linked most directly because they share the common element of "intrinsic motivation."
    a) Requires bridging inference relating cognitive conflict to students' disagreeing
    b) Requires bridging inference relating learned helplessness to attributions.
    c) Unrelated ideas.

17. **a)** The neuron is just an analogy in this case, not a component of the model of thinking.

18. **d)** hierarchical organization of ideas; thinking about one leads to thinking about the next

19. **b)**

20. **a)** propositions

Completion

1. nodes
2. macropropositions
3. hierarchical
4. ACT*
5. thematic, (as opposed to hierarchical)
6. micropropositions
7. practice
8. spreading activation
9. distributed (better than massed practice)
10. neural networks, or connectionist model

Matching

__a__ 1. Images
__c__ 2. Macropropositions
__g__ 3. Neural networks
__m__ 4. Nodes
__h__ 5. Productions
__b__ 6. Propositions
__f__ 7. Propositional networks
__e__ 8. Schemata
__l__ 9. Semantic
__d__ 10. Semantic networks

Correct terms for responses left over:
i) prototype theory
j) episodic
k) micropropositions

CHAPTER 4

Strategies and Metacognitive Regulation of Strategies

LEARNING OBJECTIVES

1. Describe the development of memory strategies in children.

2. Explain how several strategies interventions work, based on theories of knowledge representation (presented in Chapter 3).

3. Evaluate some familiar study strategies based on information from the chapter.

4. Discuss how strategies can be taught and why not all strategies can be taught to all students.

5. Explain the role of metacognition, including self-monitoring and self-regulating, in learning to apply strategies independently.

6. Discuss the role of memory in strategies use, including the implications for educational interventions.

7. Synthesize the information in the chapter to identify instructional implications.

STRENGTHENING WHAT YOU KNOW

The purpose of this chapter is to describe strategies that children use and can learn to use to help them learn information. The chapter includes information about strategies instruction, including how to encourage students to apply the strategies they have learned to new situations.

Objective 1.   **Describe the development of memory strategies in children.**
     A.   *Memory Strategies*

1. Define "strategy" in a way you will understand. Be sure to include at least 3 key features of strategies that were presented in the chapter.

2. Are strategies automatic, intentional, or both? Explain.

3. Define the following types of memory strategies and give your own example of each.

Strategy:	Definition:	Example:
Rehearsal		
Organization		
Elaboration		

4. According to the chapter, which type of strategy is least effective?

B. *Preschoolers' Memory Strategies*

1. What kind of strategies do preschoolers usually use to help them remember? (Overall, how can preschoolers' strategies be characterized)?

2. Give examples of strategies used by preschoolers.

3. Explain how preschoolers' choice of strategies can actually decrease their recall.

4. What can teachers do to encourage optimal memory performance in preschoolers?

C. *Grade-School Children's Memory Strategies*

1. What new kinds of strategies do children in the later elementary grades use to help them remember?

2. How are these strategies different than the ones preschoolers use?

3. During grade school, which kinds of strategies have the clearest increases? Which kinds of strategies are slower to develop?

D.   *Continuing Development of Strategies in Adolescents*

1. What new kind of strategies do high school students and adults use to help them remember?

2. How are these strategies different than those used by younger students?

3. What is "fast finish," and how does it enhance the cumulative rehearsal strategy?

3. Explain how cultural factors might influence the development of a child's strategies.

Objective 2.   **Explain how several strategies interventions work, based on theories of knowledge representation (presented in chapter 3).**

Complete the following table to illustrate the connection between knowledge representation theories and the strategies generated from them:

Theory of Knowledge Representation	Strategy for Understanding and Remembering

A.   *Summarization Strategies*

1. What key difference between good readers and novice readers supports the teaching of summarization strategies?

2. List Brown and Day's summarization rules:

1)

2)

3)

4)

5)

6)

3. How did Brown and Day come up with these rules?

4. Define super-ordinate. (Check a dictionary if necessary.)

5. Describe Taylor's approach to summarization training.

6. Describe Berkowitz's approach to summarization training.

7. Choose a section of the chapter to summarize using Brown & Day's approach, another to summarize using Taylor's approach, and a third to summarize using Berkowitz's approach. [Construct your summaries on separate pieces of paper.] Which strategy did you like best?

Which one do you think helped you understand the information the best?

Which one do you think will help you remember the most?

B.   *Story Grammar Training*

1. What key difference between good readers and novice readers supports the teaching of story grammar?

2. List Short and Ryan's self-questions for identifying story grammars.

    1)

    2)

    3)

    4)

    5)

3. Describe another way (other than self-questioning) that students have been taught to use story grammar.

4. How can text schemata instruction help learning of expository text, such as a content area textbook?

C. *Mental Imagery*

1. What key difference between good readers and novice readers supports instruction in mental imagery?

2. Define the following 2 kinds of imagery in a way that makes sense to you:

Representational Imagery	Transformational imagery

3. How will you remember the difference between representational and transformational images?

4. Describe the keyword method.

5. Select a vocabulary term from this chapter or an earlier chapter and describe how you could apply the keyword method to remember that term.

6. Try to generate several of your own examples, other than vocabulary learning, when the keyword method might be helpful.

7. Define the following terms:

Mimetic Reconstruction	Symbolic Reconstruction

8. How will you remember the difference between mimetic and symbolic reconstruction?

Objective 3. **Evaluate some familiar study strategies based on information from the chapter.**

    A.   *Notetaking*

1. Identify and explain the 2 primary functions of notetaking.

2. What processes make notetaking most effective?

3. When can notetaking actually hurt more than it helps?

4. In the table, write descriptions of the 3 types of notetaking mentioned in the chapter:

Conventional Format	Outline Format	Matrix Format

B.   *Outlining*

1. What is the key function of outlining?

2. Choose a section of the chapter, and create a traditional outline for it. Choose another section and create a 2-dimensional outline.

    [Create your outlines on separate paper.]

Which outline format did you like best?

Which do you think helped you understand the information the best?

Which do you think will help you remember the most?

3. When might it be ineffective to ask students to outline?

C.   *Self-questioning*

1. List several ways self-questions can improve learning for text (i.e., kinds of self-questions that are helpful).

2. What kinds of students may benefit most from self-questioning? Why do you think that is?

D.   *Rehearsal*

1. Why is rehearsal considered less effective than other strategies?

E.   *Highlighting/Underlining*

1. Why is underlining/highlighting considered less effective than other strategies?

2. What can make underlining a bit more effective?

F.   *SQ3R / PQ4R*

1. What does SQ3R stand for?

   What does PQ4R stand for?

2. Explain the steps of PQ4R.

3. What makes PQ4R more effective than SQ3R?

Objective 4. **Discuss how strategies can be taught and why not all strategies can be taught to all students.**

    A.   *Teaching Students Strategies*

1. According to the chapter, how hard is it, in general, to teach a student how to use memory strategies, such as rehearsal and organization?

2. Explain challenges in teaching mental imagery strategies to preschool children.

3. Explain challenges in teaching mental imagery strategies to grade-school children.

4. Define maintenance. How does it relate to learning a strategy?

5. Define transfer. How does it relate to learning a strategy?

6. Give some key reasons students often fail to transfer strategies. For each one, list what a teacher might do, if anything, to overcome this obstacle to strategies transfer.

Obstacle to Strategy Transfer	Instructional Implication

Objective 5. **Explain the role of metacognition, including self-monitoring and self-regulating, in learning to apply strategies independently.**

A. *Metacognitive Knowledge*

1. Write definitions in the following diagram:

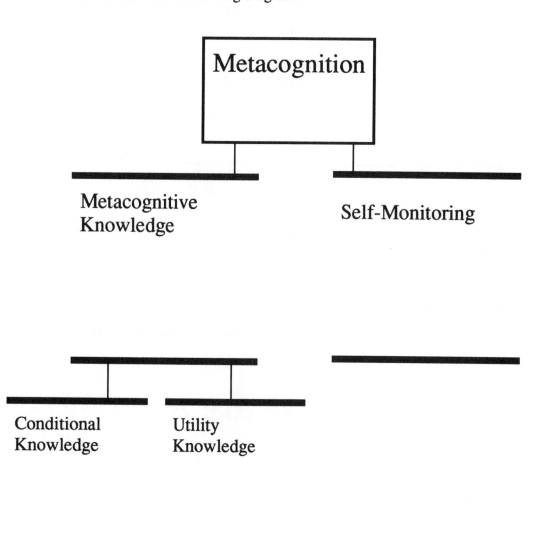

2. Describe the relationship between self-monitoring of strategies use and metacognitive knowledge about strategies.

3. What kind of metacognitive knowledge most directly affects strategy transfer?

4. What kind of metacognitive knowledge most directly affects strategy maintenance?

      B.   *Self-control training*
1. What are the key features of self-control training?

2. Describe the instructional components of self-control training.

C.   *Encouraging self-monitoring*

1. Fill in the chart to describe self-monitoring of the keyword method at different ages.

	Upper Elementary	Young Adults	Older Adults
How well did they self-monitor after studying but before testing?			
Which strategy worked better, keyword or rehearsal?			
How well did they self-monitor after testing?			
Which strategy did they choose for future learning?			
Could experimenters convince them the other strategy was more effective than keyword?			

2. Explain what was needed for upper elementary children to be <u>confident</u> that keyword was better than other methods.

3. What steps were needed for lower elementary children to chose the more effective strategy in future learning?

4. What two rationales did some older adults give for choosing a less effective strategy?
    a)

    b)

Objective 6.   **Discuss the role of memory in strategies use, including the implications for educational interventions.**

    A.   *Long-Term Memory*

1. What view of knowledge representation probably best describes how strategies are stored in long-term memory? Briefly explain the theory as it applies to strategies.

    B.   *Short-Term Memory*

1. Define short-term memory.

2. Explain how short-term memory capacity limitations can limit students' ability to use a strategy.

3. Based on this information, what kinds of students may benefit most from instruction of complex strategies? Who may benefit least?

4. Explain the impact capacity-demanding strategies can have on learner motivation.

5. Create a flow-chart depicting the memory processes involved in strategy execution (see your textbook).

6. Using the view that strategies become part of procedural knowledge, explain how strategies are learned.

7. Summarize what teachers need to know about the role of short-term memory when they are designing strategy instruction.

Objective 7. Synthesize the information in the chapter to identify instructional implications.

1. The following table lists several prerequisites for using a strategy. Explain how use of strategies could be impaired if this requirement is not met, and give an example. Suggest what a teacher might do if faced with this obstacle.

Prerequisite	Obstacles if Prerequisite Is Not Met	What Teacher Could Do if Faced With this Obstacle
Strategy procedure is in long-term memory.		
Learner has utility knowledge about the strategy.		
Learner has conditional knowledge about the strategy.		
Learner has declarative knowledge to use with the strategy.		
Learner has sufficient short-term memory capacity.		
Learner can create mental images.		
Learner is motivated to use the strategy.		

2. Across the research on strategy effectiveness, what features make a strategy more effective?

   less effective?

PRACTICE TESTS

[See answer key at the end of the chapter for correct responses.]

Multiple Choice

Circle the letter of the best response to each question.

1. Which of the following is true of strategies, according to the chapter?
    a) Learners have only unconscious regulation of strategies.
    b) Strategies are the processes needed to carry out a task.
    c) Strategies usually start working automatically when they are needed.
    d) Strategies are cognitive operations.

2. Which of the following memory strategies is probably least effective?
    a) Organization
    b) Verbal elaboration
    c) Visual elaboration
    d) Rehearsal

3. You introduce four college friends--Pam, Frank, Sarah, and Joe--to a preschooler and tell her she needs to remember their names. What is she most likely to do?
    a) Imagine each person with an object that rhymes with his or her name (such as imagining Pam eating a can of Spam).
    b) Stare at each person and say his or her name.
    c) As you introduce each person, say the names of everyone introduced so far (Pam. Pam, Frank. Pam, Frank, Sarah. Pam, Frank, Sarah, Joe.)
    d) Separate the names into girl's names and boy's names before trying to remember them.

4. What is the most likely reason preschoolers may be unsuccessful at using elaborations?
    a) Making associations between very different ideas, such as an animal and a game, is too abstract for preschoolers.
    b) It is hard for preschoolers to represent visual images mentally.
    c) The elaboration actually distracts the preschooler. For example, once the child imagines a skunk jumping rope, he might create an entire story about the animals in the woods playing games, forgetting about the original information.
    d) They do not automatically think about the elaboration when trying to remember the information.

97

5. Which of the following teaching practices would probably be most helpful to preschoolers?
   a) When presenting information, encourage students to practice by rehearsing information, rather than just saying it once or looking at it once.
   b) When testing students about something, remind them to think about how they learned it.
   c) When testing students about something, allow them to draw a picture if they can't think of how to say it.
   d) When presenting information, have students group items into meaningful categories.

6. On the first day of classes, teachers sometimes have students get to know one another through a name memory game, in which students sit in a circle and introduce themselves after saying the names of all the previous students. Which strategy does this activity most encourage?
   a) Verbal elaboration
   b) Labeling
   c) Cumulative rehearsal
   d) Encoding

7. Which of the following memory strategies is an upper elementary student least likely to use?
   a) Encoding
   b) Cumulative rehearsal
   c) Organizing
   d) Elaborating

8. Which of the following statements about strategies development is true?
   a) By college age, most students have well-developed strategies for remembering information.
   b) Preschoolers are unable to learn elaborations that associate two objects.
   c) Western schooling makes an impact on strategies development, but within Western cultures, strategies development is essentially the same.
   d) Elaboration is harder for elementary students, mainly because they lack the necessary declarative knowledge.

9. Which of the following is **not** part of Brown and Day's summarization rules?
   a) Delete trivial information.
   b) Substitute subordinate terms.
   c) Integrate a series of events with an action term that encompasses them.
   d) Select or invent a topic sentence.

10. Use of story grammar is most supported by which of the following knowledge representation theories?
   a) concept learning
   b) propositions
   c) schema theory
   d) dual-coding theory

11. Summarization is most supported by which of the following knowledge representation theories?
   a) concept learning
   b) propositions
   c) schema theory
   d) dual-coding theory

12. Mental imagery is most supported by which of the following knowledge representation theories?
   a) concept learning
   b) propositions
   c) schema theory
   d) dual-coding theory

13. Which of the following strategies is **not** directly supported by schema theory?
   a) Making a map of elements in the story.
   b) Asking oneself questions about setting, characters, goals, and outcomes.
   c) Summarizing a textbook based on a problem/solution structure.
   d) Imagining the main character in the setting solving the problem.

14. Which of the following is a representational image?
   a) As you read a scientific explanation about weather processes, you imagine each of the steps described, such as cloud formations, precipitation, and movement of the weather system.
   b) To remember that the Spanish word "pato" means "duck," you imagine a cartoon duck wearing a cooking pot as a hat; the cooking pot represents the work pato.
   c) You want to remember that your new professor is Dr. Taylor, so you imagine him with a needle and thread, sewing a suit like a tailor.
   d) You want to remember a specific event that people in a certain culture associate with freedom. You picture the statue of liberty at that event, because for you the statue of liberty represents freedom.

15. Which of the following notetaking formats is most effective, according to research?
    a) Outline form
    b) Conventional form
    c) Matrix form
    d) Verbatim form

16. Which of the following is probably the **least** important process function of notetaking?
    a) Connections with prior knowledge
    b) Organizing information
    c) Rehearsing the information
    d) Selecting key points

17. Which of the following is true of rehearsal (including rereading or recopying) and underlining/highlighting?
    a) Research shows they make no difference in learning.
    b) Rereading, recopying, or underlining takes time away from meaningful analysis of information.
    c) Study skills courses often discourage these strategies.
    d) Highlighting and underlining is more helpful when students highlight more information.

18. Of the following common strategies, which is probably the most effective?
    a) Rereading
    b) Highlighting
    c) SQ3R
    d) PQ4R

19. Which of the following is a true statement about teaching strategies to students?
    a) If students are not using rehearsal or organizing information, it is difficult for them to learn how to do so.
    b) It is easy for preschool children to learn to use interactive images, like the keyword method.
    c) Once students have learned how to use a strategy, they will probably use it as long as the task is very similar to the one they learned on.
    d) Students often need to be told about how a strategy is helpful, even if they've already experienced using it.

20. Which of the following is **not** an aspect of metacognition?
    a) Knowing how to use a strategy.
    b) Knowing the benefits of using a strategy.
    c) Knowing when to use a strategy.
    d) Knowing whether a strategy is helping while using it on a given task.

21. Which of the following was true in studies of self-monitoring strategy effectiveness?
    a) After studying, the people could sense which strategy had worked better for them.
    b) When preparing for a second testing, upper-elementary children did not choose the effective strategy because they did not recognize that it was better.
    c) Once children knew which strategy worked better for them, they chose the better strategy.
    d) Older adults did not use the better method because the other method was easier or more familiar.

22. Which of the following obstacles to strategy instruction can be most quickly and easily overcome?
    a) The students lack the short-term memory needed to carry out the strategy.
    b) The students have difficulty constructing mental images for keyword method.
    c) The students are not aware that the strategy is helping them.
    d) The students have difficulty making connections between ideas or objects that do not have a concrete relationship.

Completion

Fill in each blank with the best fitting term from the chapter. Terms are used only once.

1. Effective readers respond to vividly descriptive text by constructing
   _____ .

2. A study strategy that includes an overview of material, self-questioning, and
   making connections to prior knowledge is _____ .

3. Knowing where and when to use a strategy is _____ knowledge.

4. Knowing that a strategy is effective is _____ knowledge.

5. A diagram depicting relationships among ideas is known as a(n) _____ .

6. _____ strategies can be difficult to teach to students.

7. The kind of metacognitive knowledge most likely to affect transfer is
   _____ knowledge.

8. When first learning a strategy, it is probably represented in long-term memory as
   _____ .

9. Strategies that have been well practiced are probably are stored in long-term
   memory is in the form of _____ .

Matching

Match the letters of the description on the right with the corresponding numbered terms on the left. Use each description only once. Some descriptions may be left over.

_____ 1. Advance organizer
_____ 2. Cumulative rehearsal
_____ 3. Encoding
_____ 4. Fast finish
_____ 5. Mimetic reconstruction
_____ 6. Representational image
_____ 7. Retrieval
_____ 8. Symbolic reconstruction
_____ 9. Transfer
_____ 10. Transformational image

a) Repeating a list over and over until, over time, you remember it
b) Remembering an interactive image that represents an abstract concept in a figurative way, then using the image to recall the concept.
c) Creating a memory that will be easy to recall
d) When remembering a list of items, saying the final items first
e) Internal depiction of material as it is stated
f) Builds on prior knowledge and introduces key ideas
g) Accessing a memory
h) Using a strategy after it has been taught, in similar situations
i) Recodes part of the message into a concrete relationship with new elements
j) Saying the items presented so far in order until all items are practiced in order
k) Remembering a concrete interactive picture that is a literal representation of a fact, such as a specific instance of the fact, then using the image to recall the more abstract fact
l) Using a strategy in a new situation

LEARNING STRATEGIES

Because this chapter is about learning strategies, develop your own strategies for remembering chapter concepts that are important or hard for you to remember.

Strategy Example #1

Strategy Example #2

Strategy Example #3

ANSWER KEY

Strengthening What You Know
Objective/Item
2 C 8.
[It might help to note that the word "mimetic" refers to mimicking or imitating or that it contains the word "mime."]

Multiple Choice

Correct answers are in bold.
Comments related to other options indicate why that response is incorrect.

1. **d)**
   b) Strategies go beyond the processes needed to carry out a task.
   c) Strategies can sometimes be automatic, but more often they are conscious and intentional.

2. **d)**

3. **b)**

4. **d)**

5. **b)**

6. **c)**
   d) Encoding is adistractor--all strategies here involve encoding information, but encoding itself is not a strategy.

7. **d)**
   a) Encoding is a distractor--all strategies here involve encoding information, but encoding itself is not a strategy.

8. **d)**

9. **b)** super-ordinate, not subordinate

10. **c)**

11. **b)**

12. **d)**

13. **d)** Visual imagery is supported more by dual coding theory than schema theory.
    c) Schemata are not just for narrative stories. The problem/solution expository structure is a type of schema.

14. **a)**
    b) keyword method, a transformational imagery strategy
    d) a transformational image, specifically an example of symbolic reconstruction

15. **c)**

16. **c)**

17. **b)**
    a) Rehearsal can be effective, but it may consume more time than it's worth. Underlining sometimes produces small improvements over doing nothing at all, but the improvements are inconsistent.

18. **d)**

19. **d)**
    a) They can learn how to, and they can use the strategies, but they often fail to apply the strategies to new situations.
    b) This ability develops during the grade school years.
    c) Actually, students often fail to maintain strategies use

20. **a)** procedural knowledge--see chapter 3
    d) Monitoring, which is a form of metacognition (see chapter 1)

21. **d)**
    a) They could not identify the better strategy until after testing.
    b) They knew it was better but required more prompting to use this information.
    c) When choosing their strategy the second time, children needed to be explicitly told to think about which strategy had worked best for them. Younger children also needed to be taught how to assess which strategy worked better and to attribute performance specifically to the strategy.

22. **c)** Can simply have students compare performance with strategy and without, being explicit about the differences (i.e., higher grade).
    a) requires repeated practice until the strategy becomes automatic
    b) May need to wait until students are developmentally ready to use imagery.
    d) May need to wait until students can think more abstractly.

Completion

1. representational images
2. PQ4R
3. conditional
4. utility
5. concept map
6. mental imagery
7. conditional knowledge (utility knowledge mainly affects maintenance)
8. declarative knowledge
9. productions / procedural knowledge

Matching

__f__ 1. Advance organizer
__j__ 2. Cumulative rehearsal
__c__ 3. Encoding
__d__ 4. Fast finish
__k__ 5. Mimetic reconstruction
__e__ 6. Representational image
__g__ 7. Retrieval
__b__ 8. Symbolic reconstruction
__l__ 9. Transfer
__i__ 10. Transformational image

<u>Correct terms for responses left over</u>:
a) rehearsal
h) maintenance

CHAPTER 5

The Role of Knowledge in Thinking

LEARNING OBJECTIVES

1. Explain the role of domain knowledge in task performance in the domain.

2. Contrast the thinking of experts and novices in a domain.

3. Contrast the thinking of expert teachers and novice teachers.

4. Discuss how expertise in a domain develops.

5. Explain the relationship between knowledge and strategies.

6. Explain how knowledge can interfere with new learning.

7. Discuss instructional practices that can foster learners' expertise.

STRENGTHENING WHAT YOU KNOW

The purpose of this chapter is to show how powerful knowledge can be in determining performance and learning in various domains. The chapter compares the thinking processes of experts and novices in a variety of fields. The chapter also explores relationships between knowledge and strategies.

Objective 1.   **Explain the role of domain knowledge in task performance in the domain.**

     A.   *Expert versus Novice Performance*

1. List various ways that experts outperform novices in learning activities (e.g., experts remember more about a text on the topic than do novices).

2. Experts make different kinds of mistakes than novices do. Explain the difference.

3. What other factors can affect performance?
Discuss the relative impact of these factors.

4. Explain why child experts in a domain can remember more than adult novices.

Objective 2.   **Contrast the thinking of experts and novices in a domain.**

Fill in the following table with descriptions in your own words that show key distinctions in the thought processes of experts and novices in a topic:

Experts	Novices

Objective 3.   **Contrast the thinking of expert teachers and novice teachers.**

1. Fill in distinctions in how expert/novice teachers process classroom videos:

Expert Teachers	Novice Teachers

2. Fill in distinctions between expert and novice lesson planning:

Expert Teachers	Novice Teachers

3. Using the "Building Your Expertise" section of the chapter, fill in the concept map with information about knowledge representations of expert teachers:

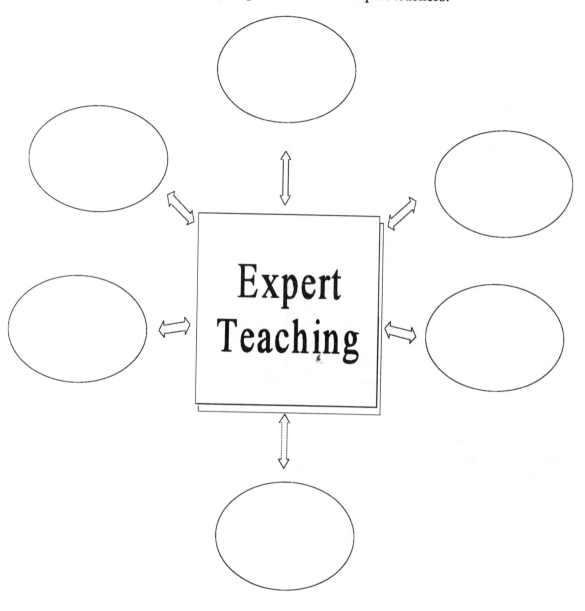

Objective 4. **Discuss how expertise in a domain develops.**

List at least 7 key factors that Bloom found influenced the development of talent:

Objective 5. **Explain the relationship between knowledge and strategies.**

    A. *Knowledge can replace strategies.*

1. Give an example of a situation where people use strategies because they lack knowledge.

2. In the situation above, what happens to strategies use once the person acquires the needed knowledge?

    B. *Knowledge can enable strategies.*

1. Give an example of a strategy that cannot be done without content knowledge.

2. For the example above, explain how knowledge interacts with the strategy.

Objective 6.   **Explain how knowledge can interfere with new learning.**
1. What kind of knowledge is most likely to interfere with new learning?

2. Explain how this kind of knowledge can get in the way of new learning.

3. Explain why even a demonstration of the correct concept can may actually strengthen students' misconceptions.

Objective 7. Discuss instructional practices that can foster learners' expertise.

A. *Practice*

Why is it critical for teachers to provide practice of skills and procedures?

B. *Interest*

1. What impact does an interesting task have on a person's thought processes?

2. How can teachers nurture student interests?

C. *Self-Regulation*

1. How does self-regulated learning help students develop expertise?

2. How can teachers promote self-regulation?
(Think back to previous chapters.)

D. *Others*

Apply concepts from the rest of the chapter to develop other suggestions for teaching.
(For example, see Bloom's research on the development of expertise).

PRACTICE TESTS

[See answer key at the end of the chapter for correct responses.]

Multiple Choice

Circle the letter of the best response to each question.

1. In studies comparing young experts, young novices, older experts, and older novices, which of the following was **not** a predictor of students' performance?
    a) intelligence
    b) age
    c) prior knowledge
    d) expertise

2. Which of the following factors has the **biggest** impact on students' learning of new information?
    a) intelligence
    b) age
    c) familiarity with the topic
    d) strategies use

3. Which of the following is an accurate comparison of experts and novices in a domain?
    a) If a text presents information that is new to everyone, the novices will learn more new information than the experts.
    b) If experts and novices in one topic are given a "neutral" topic, the "experts" still do better because of the skills they have automatized.
    c) Experts can better identify contradictions in text about the topic than can novices.
    d) Novices' mistakes in recall tend to be based on details.

4. Which of the following accurately distinguishes between experts and novices?
    a) Experts show more conscious, purposeful use of strategies.
    b) Experts who have low intelligence use more complex reasoning than highly intelligent people with less knowledge of the topic.
    c) One reason novices do not solve problems as well as experts is because they try to think about too many variables at once.
    d) Given a problem at the same time, the expert would probably still be planning when the novice has stopped planning and tried to develop solutions.

5. Which of the following statements is **false**?
   a) Experts have instant solutions to many problems in their area.
   b) Novices often don't know how much or how little they understand of information they are studying.
   c) Experts don't take up as much short-term memory because their skills are automatic and their knowledge is organized in patterns and relationships.
   d) Experts can quickly figure out how to solve a problem simply by looking at the surface features.

6. Which of the following statements about expert/novice differences is **false**?
   a) Experts check if their approach is bringing them closer to a solution, and they change their approach if it isn't working.
   b) Experts make sure they have a plan before they try to solve a problem.
   c) When solving a complex problem, experts form a detailed schema right away and fit all new information into that schema.
   d) Novices use up their thinking capacity processing the big features, giving them insufficient resources to make fine distinctions.

7. Which of the following statements represents how expert teachers process a classroom video?
   a) They focus attention on part of the room that has the most activity.
   b) They offer more interpretations and evaluations of what is happening in the classroom.
   c) They judge student behavior as appropriate of inappropriate.
   d) Their suggestions focus on specific changes the teacher should make.

8. Which of the following does **not** distinguish between expert and novice teaching?
   a) Making detailed daily plans.
   b) Planning for the long-term.
   c) Making decisions about examples in advance.
   d) Ability to stick to the planned lesson.

9. Which of the following statements about expert/novice teaching is true?
   a) Novices use routines more than experts because they require more structure.
   b) Experts have stable schemata of teaching, whereas novices change their schemata from one situation to the next.
   c) Novices form mental pictures of the ideal classroom to find ways to improve their teaching.
   d) Flexibility increases with knowledge.

10. Which of the following statements about the development of expertise is most accurate?
   a) If you work hard enough, you can become an expert in a couple of years.
   b) People who got little attention from their parents looked to their area of talent as a way to get attention.
   c) A lot of highly talented people are "self-taught."
   d) Early success is important in motivating someone to seek further opportunities in learning a skill.

11. Which of the following was recommended as a way teachers can help students become experts?
   a) Give students lots of practice in the same kinds of situations.
   b) Offer students ways to do schoolwork that is based on areas of personal interest to them.
   c) Encourage students to rely on the teacher to guide their learning.
   d) Have students memorize domain-specific facts to free their short-term memory capacity.

12. According to the chapter, differences in memory for lists are most likely due to which of the following?
   a) Greater knowledge of the topic
   b) Intentional use of strategies
   c) Strategies and knowledge
   d) Intelligence

13. Which of the following is **not** supported by the chapter?
   a) Knowledge may interfere with new learning.
   b) Knowledge can make a strategy obsolete.
   c) Knowledge is often necessary to make a strategy work.
   d) Knowledge transfers automatically to enable strategies in other domains.

14. When students have misconceptions, which of the following is most likely to happen if the teacher has students activate their knowledge before learning?
   a) Students will use strategies to overcome problems in their knowledge base.
   b) Students will be ready to learn the information because they are thinking about the topic.
   c) Students will see a difference between the previous knowledge and old knowledge, and they will adapt their schemata.
   d) Students will remember less information because wrong ideas were a block to learning.

118

Completion

Fill in each blank with the best fitting term from the chapter. Terms are used only once.

1. Knowledge in a particular field is referred to as _____ .

2. Expert teachers' flexible internal representations are known as _____ of teaching situations.

3. In the _____ phase, experts develop an initial general schema that they adapt to include further information.

4. Expert teachers often report having _____ of ideal teaching situations.

Matching

Match the letters of the description on the right with the corresponding numbered terms on the left. Use each description only once. Some descriptions may be left over.

_____ 1. Mainstream academic knowledge

_____ 2. Personal/cultural knowledge

_____ 3. Popular knowledge

_____ 4. School knowledge

_____ 5. Transformational academic knowledge

a) Knowledge about a specific topic or field

b) Knowledge presented in the media

c) Traditional, Western-culture knowledge

d) Knowledge that challenges traditional knowledge

e) Knowledge presented in textbooks and academic lectures

f) Knowledge developed from experiences in family and community

LEARNING STRATEGIES

Below are examples of strategies that can help you understand major chapter concepts. Use these examples to guide your own strategies.

Strategy Example #1

Use imagery to remember key differences between experts and novices. Picture someone you admire as an expert in his or her field (academics, sports, entertainment...), and imagine that person doing activities that represent characteristics of experts. You can also get an image of somebody who is new and not so good in that area yet. Imagine this novice doing the kinds of things most novices do.

Strategy Example #2

Think about an area or skill in your life in which you have expertise (a simple, everyday skill will do). Think about how you developed that expertise, and compare your development process to the ones described in the book. Think about the role of strategies and knowledge, and see which of these applied to you at different points in developing your expertise.

ANSWER KEY

Multiple Choice

Correct answers are in bold.
Comments related to other options indicate why that response is incorrect.

1. **a)**
    b) older experts outperformed younger experts
    d) same as prior knowledge

2. **c)** expertise

3. **c)**
    b) Expertise and its advantages to learning are domain-specific; the performance does not necessarily transfer to other areas.
    d) They often "remember" information that isn't possible in the domain.

4. **b)**
    a) Not true--usually more automatic strategies in the domain of expertise
    c) Actually, it is the experts who are more likely to simultaneously consider many variables.
    d) Experts spend <u>proportionally</u> more of their time planning, but they are still faster at solving problems than novices; the expert might even be finished solving the problem when the novice is still planning. Novices spend a lot longer solving the problem, and a smaller percentage of their time is devoted to identifying plans and patterns.

5. **d)** Experts focus on principles and concepts underlying a problem, whereas novices are easily distracted by surface features.

6. **c)** Experts start with a general schema, and then they fine-tune it as they take in more information--testing their hypotheses. They don't try to force-fit new information into their schema, as novices often do. Rather, they adapt their schema to fit the situation.

7. **b)**
    a) They evenly divide their attention.
    c) Novices are more likely to judge student behavior, whereas experts explain behavior.
    d) Novices focus on specific behaviors, whereas experts suggest principles that could guide teacher decisions.

8. **a)** Both experts and novices make detailed daily plans.
    b) Experts
    c) Novices (Experts improvised)
    d) Novices are thrown off track by student questions and misunderstandings.

9. **d)**
    a) Experts use routines more because they make procedures efficient.
    c) Experts form images of ideal classroom situations.

10. **d)**
    a) False--It takes several years of study and practice to become an expert.
    b) Most talented people had parents who strongly supported them and their skill.
    c) Most experts had intensive instruction from a master teacher.

11. **b)**
    a) Give practice opportunities in diverse situations.
    c) Students should take responsibility.

12. **c)**

13. **d)**

14. **d)**

Completion

1. domain-specific
2. mental model
3. schema invocation
4. images

Matching

__c__ 1. Mainstream academic
        knowledge

__f__ 2. Personal/cultural knowledge

__b__ 3. Popular knowledge

__e__ 4. School knowledge

__d__ 5. Transformational academic
        knowledge

Correct terms for responses left over:
a) domain-specific knowledge

CHAPTER 6

Biological Factors Affecting Learning and Development

LEARNING OBJECTIVES

1. Explain the milestones of neurological development following birth.

2. Discuss potential disruptions in neurological development.

3. Discuss interventions for overcoming disruptions and fostering normal development.

4. Explain how humans are biologically prepared to acquire certain kinds of learning, and discuss biological constraints to learning.

5. Discuss biological foundations of academic competence, including potential biological explanations for failure to acquire reading and math skills.

6. Explain perceptual learning theory--how humans construct knowledge from perceptions.

7. Explore how biological foundations can inform classroom practices.

STRENGTHENING WHAT YOU KNOW

The purpose of this chapter is to explain biological foundations of human learning and to discuss their instructional implications.

Objective 1. **Explain the milestones of neurological development before birth and following birth.**

    A. *Neurological Development Before Birth*

1. Briefly explain the process of neurogenesis

2. Define the following terms
Axon:

Dendrites:

Synapses:

3. Roughly outline aspects of brain development that occur before birth and after birth.

B.   *Myelination*

1. What is myelination?

2. List at least 3 roles of myelin.

3. In what sequence and for how long does myelination progress?

4. What disease can cause loss of myelination, and what are the consequences?

C.   *Physical Growth*

1. Briefly describe the rate and timing of growth in size of the human brain.

2. What kinds of factors can have an impact on brain size?
Give examples from your life.

D.   *Development of the Frontal Lobes*

1. Explain the kinds of development that occur in the frontal lobes of the brain.

2. What changes in infant behavior over time are associated with the development of the frontal lobes?

Objective 2.   **Discuss potential disruptions in neurological development.**
   and
Objective 3.   **Discuss interventions for overcoming disruptions and fostering normal development.**

1. Complete the following chart:

	Causes/ Explanation	Effects on neurology	Critical period when most risk?	Prevention/ Intervention
PKU				
Lou Gehrig's disease				
Teratogens				
Malnutrition				
Anoxia				

2. Explain why stimulating environments are important for children who are have biological disorders affecting neurology.

Objective 4. **Explain how humans are biologically prepared to acquire certain kinds of learning, and discuss biological constraints to learning.**

A.   *Experience-Expectant versus Experience-Dependent Synapses*

1. In the chart below, note some key differences between experience-expectant and experience-dependent synapses:

	Experience-Expectant Synapses	Experience-Dependent Synapses
Are they genetically programmed to expect certain forms of experience?		
What happens if a specific kind of stimulation is not received? How would these synapses react given a some other kind of environmental stimulus?		
When are these synapses formed?		

2. Explain cell assemblies and how they are formed.

B. *Sensation Preferences*

1. Note some of the visual preferences humans have.

2. Think about differences in toys, games, and illustrations developed for children of various ages. In what way, if any, do these changes support human's developmental biases for visual perception?

C. *Language Acquisition*

1. Explain the concept behind Chomsky's "language acquisition device."

2. What evidence supports or refutes Chomsky's claim that language is uniquely human?

3. What evidence in the chapter supports or refutes that their is a critical period for learning a second language?

D. *Hemispheric Specialization [Left-Brain / Right-Brain]*

1. Fill in the following chart, using the terms listed below it.

Functions of the Left Hemisphere	Functions of the Right Hemisphere

Time information   Language   Logical   Intuitive   Analytical
Rational   Analogies   Music   Art   Math
Spatial relationships   Sequential information   Simultaneous processing

2. A school guidance counselor suggests that students be given physical tests to see whether they are right-brain or left-brain dominant, and that this information be used to suggest possible careers or coursework.

    A) What kinds of physical tests could be given to see which side of a student's brain is dominant?

    B) Assuming that the test results are already available, what advice would you give about using the results to guide students' course choices and career choices? Support your opinion with information from the text.

131

E. *Genetically-Determined Modules in the Mind*

1. Define "cognitive modules."

2. What evidence supports the view of cognitive modules?

3. Briefly explain the theory of multiple intelligences and list the 7 intelligences.

4. A school where you are teaching is holding a faculty meeting to consider developing a curriculum based on multiple intelligences. List some comments and questions you would bring to the meeting. Support the value of your comments and questions with information from the chapter.

F.   *Capacity Constraints/Attention Span*

1. What are some explanations of how a child's attention span increases with age?

2. List some possible biological explanations for attentional deficit - hyperactivity disorder (ADHD)?

3. List potential treatments for ADHD.

4. List instructional approaches teachers can use when working with students who have short attention spans.

Objective 5. **Discuss biological foundations of academic competence, including potential biological explanations for failure to acquire reading and math skills.**

    A. *Role of Biological Factors in Reading*

Brain imaging studies show electrical activity differences for different reading processes. Summarize this information in the chart below by filling in the which part of the brain is activated in each situation. Also note possible implications or reasons for the results.

READING SITUATION	PARTS OF BRAIN ACTIVATED & Implications
Primary-grades students comprehending challenging text	
Middle-grades students comprehending challenging texts	
Reading real words	
Reading nonsense words	
Reading familiar words	

B. *Dyslexia*

1. What percentage of the population is dyslexic?

2. Define these categories and subcategories of dyslexia, listing symptoms:

a) acquired dyslexia

1) acquired phonological dyslexia

2) acquired surface dyslexia

3) acquired deep dyslexia

b) developmental dyslexia

C. *Disorders in Mathematical Competency*

1. List potential symptoms of acquired discalculia

2. List potential symptoms of developmental discalculia

3. Fill in the following table to indicate areas of the brain associated with each of the following mathematical abilities.

Mathematical Ability	Area of the Brain
Understanding word problems, math concepts, and math procedures	
Quick mental calculation, abstract conceptualizing, some problem solving skills	
Auditory processing of numbers	
Visual processing of mathematical symbols	

Objective 6.   **Explain perceptual learning theory--how humans construct knowledge from perceptions.**

1. Explain each of these basic mechanisms of perceptual learning

     a) abstraction

     b) filtering

     c) peripheral attention

PRACTICE TESTS

[See answer key at the end of the chapter for correct responses.]

Multiple Choice

Circle the letter of the best response to each question.

1. Which of the following statements accurately describes neural development?
   a) During the early elementary years, neurons multiply through normal cell division.
   c) The number of neurons stays stable throughout a person's life.
   d) Damaged neurons can regenerate themselves.

2. Which of the following conduct neural impulses away from a cell?
   a) axons
   b) synapses
   c) dendrites
   d) cell assemblies

3. Which of the following is a conduit through which nerve cells send transmissions to other cells?
   a) axon
   b) synapse
   c) dendrite
   d) cell assembly

4. Which of the following is **not** a characteristic of neural maturation after birth?
   a) Axons get a layer of sheathing.
   b) Dendrites grow in length.
   c) Additional neurons are formed.
   d) Dendrites develop more branches

5. Which of the following is **not** a critical function of myelin?
   a) Permits more rapid firing of axons
   b) Reduces recovery time needed between neural firings
   c) Insulates nerve fibers
   d) Lubricates synaptic connections

6.   Which of the following is true of experience-dependent synapses?
   a) They are genetically programmed to deal with specific kinds of information.
   b) They develop during a critical period.
   c) They are likely to die if they do not make strong connections.
   d) They can stablize to accept whatever stimuli the person encounters.

7.   Which of the following disorders can cause a loss of myelin?
   a) PKU
   b) Lou Gerhig's disease
   c) Anoxia
   d) Teratogens

8.   Fetal alcohol syndrome is an example of which of the following disorders?
   a) PKU
   b) Lou Gerhig's disease
   c) Anoxia
   d) Teratogens

9.   Which of the following represents a mothers' disease during pregnancy?
   a) PKU
   b) Lou Gerhig's disease
   c) Anoxia
   d) Teratogens

10.   Which of the following is associated with breech births?
   a) PKU
   b) Lou Gerhig's disease
   c) Anoxia
   d) Teratogens

11.   Which of the following is typically treated with a strictly controlled diet?
   a) PKU
   b) Lou Gerhig's disease
   c) Anoxia
   d) Teratogens

12.   Which of the following occurs due to an injury?
   a) PKU
   b) Lou Gerhig's disease
   c) Anoxia
   d) Teratogens

13. Which of the following is **not** directly supported by the sensation preference information discussed in the chapter?
   a) Humans are genetically programed to prefer certain colors over others.
   b) As humans develop, they prefer to look at more complex shapes.
   c) People of all ages tend to focus on the outlines of shapes.
   d) Younger infants would be more interested in a striped T-shirt, whereas older infants would rather look at a T-shirt with a bullseye.

14. Which scientist is known for theorizing that human brains include a language acquisition device that genetically prepares them to learn language?
   a) Piaget
   b) Vygotsky
   c) Chomsky
   d) Bruner

15. Which of these claims is consistent with research in genetics?
   a) Animals other than humans cannot learn to count.
   b) The capacity to create grammatical rules is unique to humans.
   c) Humans' biological preparedness for certain types of learning is limited to basic perceptions, like vision.
   d) Genetic characteristics make developing and learning language more likely for humans than for other species.

16. Which of the following statements about second language learning is true?
   a) Children learn pronunciation in a second language faster than adults do.
   b) Adults learn second language grammar more quickly than children do.
   c) Language proficiency is more certain if acquisition begins after puberty.
   d) People who immigrated to another country as adolescents were more competent in the new language than those who immigrated as children.

17. Which of the following is a function of the right hemisphere of the brain?
   a) Language
   b) Analytical information
   c) Spatial relationships
   d) Temporal relationships

18. Which of the following statements about left-brain/right-brain theory is **not** yet supported by research?
    a) Blood flows more to a particular side of the brain for English majors, but to the other side for architects.
    b) Distinct eye movements are associated with each side of the brain.
    c) Individual differences in hemispheric specialization are stable over time.
    d) Differences in hemispheric functioning are associated with particular talents.

19. Increases in attention span/executive processing space are associated with increases in all of the following **except** which one?
    a) Automazation
    b) Neurological maturation
    c) Practice
    d) Effort

20. Which of the following is a symptom of acquired deep dyslexia?
    a) Saying "bird" when you see the word "pigeon"
    b) Mistaking right for left
    c) Reading better when the book is upside-down
    d) Abnormal eye-scan patterns

21. Which of the following is **not** an early warning sign of dyslexia?
    a) Short attention span
    b) Poor pronunciation
    c) Difficulty naming common objects
    d) Using relatively short sentences

22. Which of the following is supported as a neurological explanation for dyslexia?
    a) Under-developed left hemisphere
    b) Under-developed right hemisphere
    c) Dysfunctional eye movements
    d) Abnormal brain tissue

23. Which of the following perceptual learning mechanisms involves finding constant relationships among a class of events?
    a) Abstraction
    b) Filtering
    c) Modulation
    d) Peripheral attention

24. Which of the following perceptual learning mechanisms guides eye movement to important information?
   a) Abstraction
   b) Filtering
   c) Modulation
   d) Peripheral attention

25. Which of the following perceptual learning mechanisms involves disregarding unimportant information?
   a) Abstraction
   b) Filtering
   c) Modulation
   d) Peripheral attention

Completion

Fill in each blank with the best fitting term from the chapter. Terms are used only once.

1.   The life of _____ synapses depends on whether appropriate environmental stimuli are present during a critical period.

2.   _____ synapses adapt to whatever environmental stimulation the individual encounters.

3.   Chomsky believes human brains have a genetically-determined _____ device.

4.   A reading disorder resulting from a brain injury is called _____ _____.

5.   More common than brain injury cases, the neurological reading disorder educators most often encounter is _____ _____.

6.   A student who has difficulty reading or writing mathematical symbols or who has trouble calculating math problems may have _____ _____.

7.   _____'s theory of multiple intelligences suggests that people vary in their strength in several biologically-determined domains.

8.   In perceptual learning, young children's attention is often captured by _____ environmental stimuli.

9.   Gibson believes that although people are biologically prepared to get certain kinds of information from the environment, learning is gradual and varies with _____.

Matching

Match the letters of the description on the right with the corresponding numbered terms on the left. Use each description only once. Some descriptions may be left over.

_____ 1. Anoxia
_____ 2. Axon
_____ 3. Cell assembly
_____ 4. Cerebral cortex
_____ 5. Cognitive module
_____ 6. Dendrite
_____ 7. Myelin
_____ 8. Neurogenesis
_____ 9. Synapses
_____ 10. Teratogen

a) Section of the brain responsible for many complex thought processes
b) Sheathing that permits rapid firing of neural impulses
c) Chemical-related or disease-related agent that can cause prenatal brain damage
d) Branch-like extension that transmits impulses toward cell body from other cells
e) Reduced air supply during birth
f) Reading disorder
g) Period of rapid cell reproduction through division
h) Mathematical disorder
i) A system genetically designed for a specific mental function
j) Conducts impulses away from cell body
k) Closed path of several connected neurons
m) Connections between neurons

LEARNING STRATEGIES

Below are examples of strategies that can help you understand major chapter concepts. Use these examples to guide your own strategies.

Strategy Example #1
Draw a diagram that shows relationships among the following:
Nerve cells   Axons   Dendrites   Synapses   Myelin

Strategy Example #2
Try to apply the keyword method (chapter 3) to any vocabulary you are having difficulty remembering.

Strategy Example #3
Role-play: Pretend you are the principle of a school where teachers are discouraged by students' biological constraints to learning. Write a motivational speech that acknowledges their specific concerns and suggests ways they can cope with students' challenges.

ANSWER KEY

Multiple Choice

Correct answers are in bold.
Comments related to other options indicate why that response is incorrect.

1. **b)**
   a) False--Rapid cell growth occurs before birth. After this neurogenesis is completed, there will be no new neurons
   c) Although no new neurons will form, neurons die--often in the first few years of life; elimination of excess nerve cells and synapses is considered necessary to cognitive development.
   d) Neurons lost due to disease or injury will not be replaced.

2. **a)**
   c) Dendrites transmit neural impulses to a cell from other cells.

3. **b)**
   a) The transmission process requires not just an axon, but synaptic connections between axons, dendrites, and cell bodies.

4. **c)** All the neurons are already formed.

5. **d)**

6. **d)**
   a, b, and c are true of experience-expectant synapses.

7. **b)**

8. **d)**

9. **d)**

10. **c)**

11. **a)**

12. **c)**

13. **a)**

14. **c)**

15. **d)**

16. **b)**
    c)   before puberty

17. **c)**

18. **c)**

19. **d)**

20. **a)**   symptom of acquired deep dyslexia

21. **a)**

22. **d)**
    a)   Activity in the left hemisphere is less predictable in dyslexics than non-dislexics. However, there is currently no evidence that the left hemisphere is under-developed

23. **a)**
    c)   distractor (not a term presented in the chapter)

24. **d)**
    c)   distractor (not a term presented in the chapter)

25. **b)**
    c)   distractor (not a term presented in the chapter)

Completion

1. experience-expectant
2. experience-dependent
3. language acquisition
4. acquired dyslexia
5. developmental dyslexia
6. developmental dyscalculia
7. Gardner
8. irrelevant
9. experience

Matching

__e__   1. Anoxia
__j__   2. Axon
__k__   3. Cell assembly
__a__   4. Cerebral cortex
__i__   5. Cognitive module
__d__   6. Dendrite
__b__   7. Myelin
__g__   8. Neurogenesis
__m__   9. Synapses
__c__   10. Teratogen

<u>Correct terms for responses left over</u>:
f) dyslexia
h) discalculia

147

CHAPTER 7

Psychological Theories of Learning and Development

LEARNING OBJECTIVES

1. Delineate Piaget's four stages of cognitive development.

2. Explain Piaget's views of the mechanisms of cognitive change.

3. Discuss the teaching implications of Piaget's theory.

4. Explain some of the criticisms of Piagetian theory.

5. Distinguish operant conditioning from classical conditioning, explaining the processes and instructional implications of each.

6. Discuss the role of behaviorism in educational settings, including the role of programmed instruction and teaching machines, behavior modification and token economies, and punishment in schools.

7. Explain Bandura's views about observational learning.

8. Describe four cognitive mechanisms that mediate social learning.

9. Compare and contrast cognitive, behavioral, and social learning theories, including their instructional implications.

10. Describe cognitive behavior modification instructional approaches.

STRENGTHENING WHAT YOU KNOW

The purpose of this chapter is to present psychological theories of learning and development, which focus on the individual learner in interactions with his or her environment. The chapter also discusses teaching implications of the theories.

Objective 1.　　　**Delineate Piaget's four stages of cognitive development.**

　A.　*Comparing the Four Stages*

1. Complete the following table:

STAGE	ages*	key accomplishments	Schemes & how acquired
sensorimotor			
preoperational			
concrete operations			
formal operations			

Be sure that your table includes the following terms and their definitions:

deferred imitation	dominated by perception	operational schemes	compensation
symbolic schemes	thinking in possibilities	object permanence	egocentrism
motor schemes	class inclusion	reflexes	reversibility	conservation
symbolic play	hypothesizing	seriation	thinking ahead	identity

*Ages are a guideline only--Piaget warned against applying age norms because development rate varies.

B. *Conservation*

1. Define conservation as it relates to Piaget's theory.

2. Explain why preschool children are often unable to solve conservation tasks.

3. Explain how each of the following operational schemes contributes to the development of conservation.

   a) identity

   b) reversibility

   c) compensation

C. *Progress Through the Stages*

1. According to Piaget, in what ways might progress through the stages vary among individuals?

2. According to Piaget, in what ways is progress through the stages universal (the same among all individuals)?

Objective 2.   **Explain Piaget's views of the mechanisms of cognitive change.**

    A.   *Key Determinants of Cognitive Development*

1. Briefly explain Piaget's views of the role of three determinants of cognitive growth:

    a) maturation

    b) experience

    c) social environment

2. Define the following terms and explain their relationship

    assimilation

    accommodation

    disequilibrium

    equilibrium

    equilibration

Objective 3.   **Discuss the teaching implications of Piaget's theory.**
   A.   *Producing Legitimate Conceptual Change*
1. According to Piaget, what is the only way instruction can produce legitimate conceptual change?

2. Based on what you have learned about Piaget, what do you think he would say happens when the teacher tries to encourage a level of thinking that is far beyond the student's current level?

   B.   *Diagnosing Students' Current Developmental Stage*
and   C.   *Appropriate Instruction at Each Stage*
1. In the chart below, list a couple of tasks that could identify whether the student is at that stage/has achieved its major accomplishments. In the far right column, give examples of appropriate materials and activities at each stage. Base your decisions on descriptions of each stage provided in the chapter.

STAGE	DIAGNOSING	INSTRUCTION
sensorimotor		
preoperational		
concrete operations		
formal operations		

2. What other general instructional principles based on Piaget apply across developmental levels?

Objective 4. **Explain some of the criticisms of Piagetian theory.**

1. List at least 6 criticisms of Piaget's theory

a)

b)

c)

d)

e)

f)

2. Summarize the author's evaluation of the usefulness of Piaget's theory in education.

Objective 5.   **Distinguish operant conditioning from classical conditioning, explaining the processes and teaching implications of each.**

    A.   *Classical Conditioning*

1. Where is the focus on learning placed in classical conditioning?
Give some examples of school situations in which this focus might apply.

2. Create a flow-chart to illustrate and explain Pavlov's experiment. Include the following terms and how they were represented in the experiment: unconditioned stimulus, unconditioned response, conditioned stimulus, conditioned response

3. Create a similar flow-chart to the one above, this time illustrating the classical conditioning explanation of how a student develops test anxiety.

4. What might a teacher do to reduce a student's test anxiety?
What is the term for this process?

B.   *Operant Conditioning*

1. Where is the focus on learning placed in operant conditioning? Give some examples of school situations in which this focus might apply.

2. Create a flow-chart to illustrate the paycheck example of operant conditioning. Include the following terms and the related examples: unconditioned reinforcer; conditioned reinforcer, extinction/extinguished

3. Create a similar flow-chart to the one above, this time illustrating a possible operant conditioning explanation for why a student disrupts class.

4. Based on the above scenario, what might a teacher do to reduce the student's disruptions? What kinds of teacher reactions would most likely lead to more disruptions? Why? [Label your suggestions with the appropriate terms.]

5. The following chart may help you distinguish the bolded terms. Include your own examples. [Note that these common examples are just to illustrate. The author neither promotes nor criticizes these practices. Keep in mind that the critical point is what is reinforcing or punishing for the individual.]

	Stimulus Is Presented (+)	Stimulus Is Removed (-)
Stimulus Is Desired	**positive reinforcement** (praise, bonus points, good grades, pizza party, stickers, money for grades)	**"removal" punishment** (time out, no TV, lose phone privileges, allowance cut)
Stimulus Is Undesired	**presentation punishment** (50 pushups, spanking, note to parent, scolding, extra homework)	**negative reinforcement** (no homework, get out of doing chores, parent stops nagging, symptoms go away when take medicine)

6. To what extent should reinforcement and punishment be used?

7. Which works better, continuous or intermittent reinforcement? Explain why.

8. Fill in the following chart with definitions and examples.

	Continuous Reinforcement Schedule	Intermittent Reinforcement Schedule	
		Ratio Schedule	Interval Schedule
Definition			
Examples			

9. In what key way can operant conditioning backfire?
What can teachers do to avoid this pitfall?

C.   *Behaviorist Vocabulary* [See strategy #3]

Complete the following chart with definitions and everyday examples of these terms:

TERM	DEFINITION	EXAMPLE
operant conditioning		
classical conditioning		
unconditioned stimulus		
conditioned stimulus		
unconditioned response		
conditioned response		
unconditioned reinforcer		
conditioned reinforcer		
ratio reinforcement		
interval reinforcement		
positive reinforcer		
negative reinforcer		
punishment		
token reinforcer		
response cost		
timeout		
extinction /extinguished		
fading		
shaping		

What is the difference between a negative reinforcer and punishment?

Objective 6.   **Discuss the role of behaviorism in educational settings, including the role of programmed instruction and teaching machines, behavior modification and token economies, and punishment in schools.**

    A.   *Programmed Instruction*

1. Describe the basic principles/components of programmed instruction. How does this keep motivation high?

2. Programmed instruction as it was originally developed is no longer widely used. What other form of instruction has developed from programmed instruction?

    B.   *Behavior Modification and Token Economies*

1. Describe the key principles behind behavior modification.

2. In what educational settings are these approaches most widely used?

3. List a range of types of learning that can be supported by behavior modification approaches.

4. Give an example of token reinforcement

5. What is a drawback of implementing a token economy? How is this usually solved?

6. What is meant by response cost?

159

7. What do teachers model in effective behavior modification programs?

8. When do effective behavior modification teachers give attention or not give attention?

9. Describe behavior contracting.

10. Explain how timeout works and how it is punishing.
In what way might timeout backfire? How can this be prevented?

C. *Punishment*
1. According to Skinner, when should punishment be used?

2. What should accompany punishment?

3. According to Skinner, what are some problems with punishment?
What have other researchers found about these potential problems?

4. What is a characteristic of effective punishment?

5. What was Skinner's view of corporal punishment, such as spanking?

Objective 7.   **Explain Bandura's views about observational learning.**

1. What is the main principle of social learning theory?

2. Explain the meaning of observational learning and vicarious experiences.

3. Explain the role of reinforcement and punishment in observational learning.

4. Explain the meaning of Bandura's "formula" for determining the probability of a behavior, and give an example.

Objective 8.   **Describe four cognitive mechanisms that mediate social learning.**

Summarize the role of each of the following cognitive mechanisms in social learning. How does each one affect the likelihood that a student will produce a behavior?

Attention Processes	
Retention Processes	
Motor Reproduction Processes	
Motivational Processes	

Objective 9.   Compare and contrast cognitive, behavioral, and social learning theories, including their instructional implications.

1. In the margin next to each statement, put a letter showing which of the theories supports the statement. [Some statements will apply to more than one theory.]

        C = Cognitive
        B = Behavioral
        S = Social Learning

People learn through interactions with the environment.
Appropriate experiences allow realization of potential provided by genetic inheritance.
Imitation is innate.
The theory emphasizes the importance of biological maturation.
Seeing somebody else being reinforced or punished can affect my behavior.
The learner has an active role.
Each stage of development a prerequisite for the next.
Learning occurs mainly through patterns of reinforcement and punishment.
What you will learn is affected by the kinds of people around you.
Focuses on observable actions rather than internal processes.
A developmental sequence is universal.
We learn a lot of behaviors without actually getting reinforced.
Theory is concerned with mental structure.
Imagining yourself writing a fabulous essay might help you complete the essay.
Theory includes learning through vicarious experiences.
Not everyone will look to the same person as a role model.
The impact of external factors is affected by what each individual likes and dislikes.
True conceptual change can only be guided through processes that mirror natural
      development.
Other people serve as behavioral models.

Objective 10.   Describe cognitive behavior modification instructional approaches.

    A.   *Self-instruction*

1. Define the key component of cognitive behavior modification.

2. Draw a flow-chart showing the process teachers can follow to teach students to use self-instructions:

PRACTICE TESTS

Multiple Choice

Circle the letter of the best response to each question.

1. Which of the following is a characteristic of Piaget's stage theory view of cognitive development?
   a) The sequence of the stages varies across individuals.
   b) Knowing a child's age will tell you his or her cognitive stage.
   c) People at higher stages have more ability to learn from the environment.
   d) People get more knowledge, but the type of knowledge is essentially the same.

2. Which of the following is a key accomplishment of the concrete operations stage?
   a) symbolic play
   b) object permanence
   c) hypothesizing
   d) conservation

3. A child is sorting pictures of her classmates so that they are in alphabetical order, with pictures of the same person organized by year the photo was taken. What stage does this behavior best represent?
   a) concrete operations
   b) formal operations
   c) preoperational
   d) sensorimotor

4. Which of the following types of schemes is associated with the preoperational stage?
   a) symbolic schemes
   b) operational schemes
   c) motor schemes
   d) possibility schemes

5. Which of the following statements best represents the sensorimotor stage?
   a) By the end of the sensorimotor stage, children love to play "peek-a-boo."
   b) Children in this stage like to point at objects around the room and name them.
   c) A child in this stage could sort a set of blocks from short to tall.
   d) A child who puts a new toy in his mouth is displaying intelligence.

6. Which of the following is **not** one of Piaget's key determinants of cognitive growth?
   a) social environment
   b) conservation
   c) biological maturation
   d) experience

7. A student adds a positive and negative number using a number line, and she gets an answer that is less than one of the addends. This violates her mental rule that when you add two numbers, the sum is greater than either of the addends. She knows she is using the number line correctly, yet she also believes in her rule. It doesn't make sense. This student is exhibiting _____.
   a) Assimilation
   b) Accommodation
   c) Disequilibrium
   d) Equilibration

8. Which of the following evaluations of Piaget's theory is best supported by the chapter?
   a) There are so many problems with Piaget's theory that it should not be used be when developing instruction.
   b) Piaget thought that a person who can solve physics problems at the formal operations level would also be able to solve math problems at the formal operations level, but research has shown this is not the case.
   c) Piaget's theory is valuable because, unlike other theories, it describes development of thinking across the lifespan.
   d) Contrary to Piaget's view, ways of thinking like conservation can often be taught with simple procedures.

9. A student who is reluctant to feed the class pet may have been bitten by a similar animal years ago. The fearful reaction occurs even though the class pet is friendly. What type of learning does this best illustrate?
   a) Piagetian learning theory
   b) Operant conditioning
   c) Social learning
   d) Classical conditioning

10. Which of the following reinforcement schedules often leads to a decline in performance right after the reinforcement?
   a) continuous reinforcement schedule
   b) intermittent reinforcement schedule
   c) ratio reinforcement schedule
   d) interval reinforcement schedule

11. An event following a behavior that makes it less likely to occur again is called:
   a) positive reinforcer.
   b) negative reinforcer .
   c) punishment.
   d) unconditioned reinforcer.

12. A professor has several students who earned A's on all of their chapter tests. As a reward, those students have the option of skipping the final. This illustrates:
   a) positive reinforcer
   b) negative reinforcer
   c) punishment
   d) unconditioned reinforcer

13. Which of the following is true of classical conditioning?
   a) It is a prominent part of school learning.
   b) It focuses on uncontrollable responses.
   c) It focuses on reinforcers.
   d) It focuses on intentional behavior.

14. Students are expected to answer a teacher's questions in class. Which of the following would an operant theorist be most interested in?
   a) Student feels nervous when responding in class.
   b) Student's voice shakes when responding in class.
   c) Student knows how to respond based on an understanding that teacher is discussing a hypothetical situation.
   d) Student volunteers to answer five questions by raising hand.

15. Which of the following is true of negative reinforcement?
   a) It weakens behavior.
   b) It is a form of classical conditioning.
   c) It allows the person to avoid something unpleasant.
   d) It is also called punishment.

16. Which of the following best represents how a teacher could get a student who has never sung in public to perform a solo in the school variety show?
    a) Fading
    b) Continuous reinforcement
    c) Punishment
    d) Shaping

17. A teacher is encouraging students to keep a diary of their study habits. Each time everyone in the class has turned in five entries, the class is allowed to eat snacks in class. Which of the following terms most specifically applies to this situation?
    a) continuous reinforcement schedule
    b) intermittent reinforcement schedule
    c) ratio reinforcement schedule
    d) interval reinforcement schedule

18. Which of the following behavioral teaching practices is now used mainly with special student populations?
    a) Computerized instruction
    b) Token economy
    c) Punishment
    d) Timeout

19. Select the theory that asserts that other people serve as behavioral models
    a) Cognitive Theory
    b) Behaviorism
    c) Social Learning Theory
    d) Cognitive Behavior Modification

20. Which of the following theorists believed that biology determined a universal sequence of development?
    a) Bandura
    b) Piaget
    c) Skinner
    d) none of the above

21. Which of the following theorists most emphasized the role of biology in development?
    a) Pavlov
    b) Piaget
    c) Skinner
    d) Bandura

22.   In Social Learning Theory, which of the following would **not** be emphasized as determining the probability of performing a given behavior?
      a) sophistication of one's cognitive scheme for the behavior
      b) beliefs about what happens as a result of the behavior
      c) sophistication of procedural knowledge of the behavior
      d) how much one likes or dislikes a specific reinforcer or punishment

23.   Which of the following concepts represents the key distinguishing feature of cognitive behavior modification?
      a) modeling
      b) self-reinforcement
      c) self-instruction
      d) fading

24.   Which of the following instructional approaches would Piaget have criticized?
      a) Figure out students' level of thinking and give tasks matched to their level.
      b) Teach students rules that will help them understand more complex ideas.
      c) Give students objects to manipulate, even if they have already reached formal operations.
      d) Allow students to continue arguing about how to solve a problem after it is clear that some of the students have become confused.

25. Which of the following statements is best supported by behaviorist theory?
      a) Teacher praise will increase the desired behavior.
      b) It is a good idea to let students pick their own rewards.
      c) Teacher ridicule is a negative reinforcer.
      d) Punishment is often necessary because giving students rewards only goes so far.

26.   Which of the following types of reinforcement is **not** directly addressed in Bandura's theory?
      a) personal history of reinforcement
      b) vicarious reinforcement
      c) self-reinforcement
      d) intermittent reinforcement

27. Which of the following does **not** represent a key teaching approach in cognitive behavior modification?
      a) Act out how you as adult would help yourself through the problem.
      b) Encourage students to whisper.
      c) Give students points whenever they do the desired behavior.
      d) Teach students to recognize situations when they are likely to misbehave.

Completion

Fill in each blank with the best fitting term from the chapter. Terms are used only once.

1. By repeating in public a word he heard Daddy say last month when he burned his hand while cooking, a preoperational child displays _____.

2. Acting out movie scenes with stuffed animals is an example of _____.

3. The understanding that you can make up for change in one dimension by changing another dimension is called _____.

4. Solving problems with hierarchical relationships, where one set of items is a subset of another, requires understanding of _____.

5. The fact that students do not master all problems requiring the same logical operation at the same time is called _____.

6. The Premarck principle suggests that _____ frequency behavior be contingent on _____ frequency behavior.

7. A behavioral _____ involves students in setting goals, choosing reinforcements, and determining criteria for success.

8. Getting demerits (or losing points) for unacceptable behavior is an example of _____.

Matching

Match the letters of the description on the right with the corresponding numbered terms on the left. Use each description only once. Some descriptions may be left over.

_____ 1. Accommodation
_____ 2. Assimilation
_____ 3. Behavioral model
_____ 4. Conservation
_____ 5. Disequilibrium
_____ 6. Egocentrism
_____ 7. Equilibration
_____ 8. Hypotheses
_____ 9. Motor schemes
_____ 10. Object permanence
_____ 11. Operational schemes
_____ 12. Self-instruction
_____ 13. Self-reinforcement
_____ 14. Seriation
_____ 15. Symbolic schemes
_____ 16. Vicarious reinforcement

a) Actions on the environment, based on reflexes, that guide human learning
b) Representing objects or events through language, mental images, and gestures
c) Understanding that reshaping something does not change its amount
d) Person one imitates when learning something
e) Logical rules for thinking
f) Awareness that a hidden object still exists
g) Arranging objects in order
h) Possibilities that can be tested systematically
i) Current understanding is consistent with incoming information.
j) Understanding only one's own point of view
k) Teaching yourself how to do something without the help of a teacher
l) Changing a way of thinking because of new information
m) Influence by seeing consequences another person receives for a behavior
n) Realizing a current way of thinking and contradictory information cannot both be true
o) Deciding to let yourself have a treat when you reach a goal
p) Fitting information into a way of thinking
q) Talking oneself through a challenging task

LEARNING STRATEGIES

Below are examples of strategies that can help you understand major chapter concepts.
Use these examples to guide your own strategies.

Strategy Example #1
Draw a flow-chart of the process of cognitive change, clarifying the relationship between assimilation, accommodation, disequilibrium, equilibrium, and equilibration

---

Strategy Example #2
To remember the differences in thinking at each of Piaget's developmental stages, visualize children of different ages sitting around a table eating pizza. A **sensorimotor** infant explores the pizza by rubbing it on her face and eating it; if she drops a piece under the table, she will look for it only if she has achieved object permanence. A **preoperational** preschooler complains about not having enough pizza (egocentric), but is easily appeased when the waiter cuts a slice in half and says, "Now you have two pieces." [The child believes this because he does not understand conservation of matter.] A **concrete operational** child who asks for more pizza does not fall for cutting the piece in half and saying it's more, and he argues with the waiter that it's the same amount. When the waiter serves additional slices, the child arranges them on his plate by width, but the pieces are different lengths. [Child can seriate, but only on 1 dimension.] A **formal operational** teenager calculates an algebraic formula to divide the pizza evenly. The teenager hypothesizes that supreme pizza weighs more than plain cheese, and she designs a balance to test her hypothesis. She entertains the other children with an elaborate story about how life would be different if the earth were made out of pizza.

Strategy Example #3
Students often misunderstand or forget meanings and distinctions among these terms:
unconditioned stimulus / conditioned stimulus / unconditioned response / conditioned response
unconditioned reinforcer / conditioned reinforcer / ratio reinforcement / interval reinforcement
positive reinforcer / negative reinforcer / punishment / token reinf./ response cost /timeout
extinguished / extinction / fading   operant conditioning / classical conditioning

Write definitions of them in your own words, then monitor your understanding by checking against the chapter. Once you are sure you have identified the correct meanings, design strategies for remembering the terms and distinguishing them from the other terms. [You might also want to take a learning tip from the behaviorists and give yourself a reward when you can remember all the terms correctly.]

ANSWER KEY

Strengthening What You Know

Objective/Item
9. 1.

C = Cognitive
B = Behavioral
S = Social Learning

People learn through interactions with the environment. [CBS]
Appropriate experiences allow realization of potential provided by genetic inheritance.
    [CBS]
Imitation is innate. [S]
The theory emphasizes the importance of biological maturation. [C]
Seeing somebody else being reinforced or punished can affect my behavior. [S]
The learner has an active role. [CB possibly S]
Each stage of development a prerequisite for the next. [C]
Learning occurs mainly through patterns of reinforcement and punishment. [B]
    [S--mediator]
What you will learn is affected by the kinds of people around you. [S]
Focuses on observable actions rather than internal processes. [B]
A developmental sequence is universal. [C]
We learn a lot of behaviors without actually getting reinforced. [S]
Theory is concerned with mental structure. [C]
Imagining yourself writing a fabulous essay might help you complete the essay. [S]
Theory includes learning through vicarious experiences. [S]
Not everyone will look to the same person as a role model. [S]
The impact of external factors is affected by what each individual likes and dislikes. [B,
    S--value]
True conceptual change can only be guided through processes that mirror natural
    development. [C-equilibration]
Other people serve as behavioral models. [S]

Multiple Choice

Correct answers are in bold.
Comments related to other options indicate why that response is incorrect.

1. **c)**
    a) Piaget believed the sequence was biologically inevitable.
    b) People progress through the stages at varying rates. Age norms are just a rough guideline.

2. **d)**
    a) preoperational
    b) sensorimotor
    c) formal operations

3. **b)** seriating on more than one dimension

4. **a)**
    b) concrete operations
    c) sensorimotor
    d) distractor--not a term in the chapter

5. **d)** Children in the sensorimotor stage use reflexes as their way of learning about the world.
    a) By the end of the sensorimotor stage, a child will know that the person is still there (object permanence) and will not be so interested in peek-a-boo.
    b) Ability to use symbols such as language is an accomplishment of the preoperational stage.
    c) concrete operations

6. **b)** Conservation is a type of thinking that emerges in cognitive growth, it is not a cause of development, but a result.

7. **c)**

8.   **d)** Conservation has been taught through correction, being told a rule, and observing a person who understands the concept while they model the task.
   a) There are some problems, but Piaget defined important issues, and the basic philosophy still influences education.
   b) Piaget did believe that people reach **formal** operations in different areas. However, he thought that **concrete** operational thinking applied across all tasks. Other theorists have suggested that students can be preoperational for some tasks and concrete operational for other tasks.
   c) The theory only shows a fraction of lifespan development, with little emphasis on adult development. Moreover, it does not necessarily fit across cultures.

9.   **d)**

10. **d)**

11. **c)**

12. **b)**

13. **b)**
   a, c, and d are true of operant conditioning.

14. **d)** Observable and controllable behavior.
   a) most related to classical conditioning, but even classical theorist would be more interested in behavior
   b) involuntary response--classical conditioning
   c) unobservable--of more interest to cognitive theorists

15. **c)**
   b) operant

16. **d)** Teacher would need to reinforce behavior gradually--For example, first reinforce humming a couple of bars along with other students; then reinforce small solo parts with choir in the background....

17. **c)**
   b) broader term; ratio and interval are both examples of intermittent reinforcement; ratio more specifically describes the situation

18. **b)**

19. **c)**

20. **b)**

21. **b)**

22. **a)**
    b) expectancy
    c) need to know how to do the behavior--procedural knowledge
    d) value of expected consequence

23. **c)**

24. **b)**
    c) Even though the students can think abstractly, the manipulatives give them a way to participate actively in discovery.
    d) Such disequilibrium is considered a natural and necessary part of cognitive growth.

25. **b)** Different students value different things, so what is reinforcing for one student may actually reduce behaviors of another student. Ultimately, it's important to encourage students to work for their own satisfaction.
    a) It depends on whether the student wants praise. With older students, such praise is considered embarrassing. In some social groups, students try to appear not to be succeeding, even if academic success is personally important to them.
    c) Punishment should be limited and is often unnecessary.

26. **d)** Bandura does not address schedules of reinforcement. The other three options are all sources upon which the learner bases expectations.

27. **c)** behavior modification rather than cognitive behavior modification; in cognitive behavior modification, teacher is more likely to encourage students to self-reinforce

Completion

1. deferred imitation
2. symbolic play
3. compensation
4. class inclusion
5. horizontal decalage
6. high, low (Students must do the less desirable activity before the more desirable activity.)
7. contract
8. response cost

Matching

__l__ 1. Accommodation
__p__ 2. Assimilation
__d__ 3. Behavioral model
__c__ 4. Conservation
__n__ 5. Disequilibrium
__j__ 6. Egocentrism
__i__ 7. Equilibration
__h__ 8. Hypotheses
__a__ 9. Motor schemes
__f__ 10. Object permanence
__e__ 11. Operational schemes
__q__ 12. Self-instruction
__o__ 13. Self-reinforcement
__g__ 14. Seriation
__b__ 15. Symbolic schemes
__m__ 16. Vicarious reinforcement

Correct terms for responses left over:
k) [none--distractor for self-instruction]

CHAPTER 8

Social Interactional Theories of Learning and Development

LEARNING OBJECTIVES

1. Explain Vygotsky's theory of the developmental relationship between thought and speech.

2. Explain Vygotsky's views about how sophisticated thought develops.

3. Compare and contrast Vygotsky's and Piaget's theories of cognitive development.

4. Explain Bakhtin's theory of social languages.

5. Describe apprenticeship as a way of learning, and discuss its classroom implications.

6. Describe Reciprocal Teaching.

7. Describe studies that support and refute the claims of social interaction theorists.

8. Compare and contrast Initiate-Respond-Evaluate cycles with true academic dialogues, including the impact of these instructional conversations on students.

9. Discuss factors that promote effective academic discussions.

STRENGTHENING WHAT YOU KNOW

The purpose of this chapter is to introduce theories of learning and development that emphasize the role of human interaction. The chapter also discusses instructional implications of these theories.

Objective 1.   **Explain Vygotsky's theory of the developmental relationship between thought and speech.**

This graphic illustrates Vygotsky's stages of development of the relationship between thought and speech. On the graphic, list key attributes of each stage.

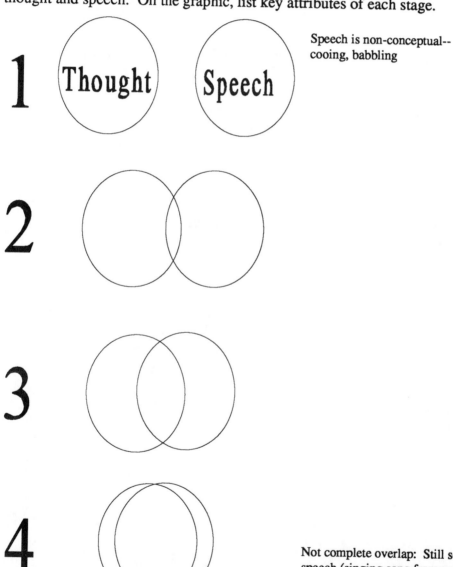

1 Thought Speech

Speech is non-conceptual--
cooing, babbling

2

3

4

Not complete overlap: Still some non-conceptual speech (singing song from memory without understanding) and non-verbal thought (skilled at using tools but can't explain how it's done).

Objective 2. Explain Vygotsky's views about how sophisticated thought develops.

    A.   *The Role of Adult/Child Interactions*

1. What role does adult assistance play in Vygotsky's theory of children's development of sophisticated thinking?

2. Explain Vygotsky's view that cognitive functions develop twice.

    B.   *Zone of Proximal Development*

Define zone of proximal development.

    C.   *Scaffolding*

Explain the concept of scaffolded instruction, showing how the analogy of a builder's scaffold relates to this kind of instruction.

Objective 3.   **Compare and contrast Vygotsky's and Piaget's theories of cognitive development.**

Compare theorists by filling in the table with information. Where a row spans across the two theorists, explain the point of similarity. Otherwise, focus on the differences.

	VYGOTSKY	PIAGET
Time of their Research		
Time of most influence		
Dialectical		
Role of individual		
Dynamic Development		
ZPD / Disequilibrium		
Role of social world		
Role of language **egocentric/ internalized speech**		

Objective 4.   Explain Bakhtin's theory of social languages.

    A.   *Social Languages / Ventriloquating / Multivoicedness*

1. Explain the concept of people using a variety of social languages.

2. How does this idea relate to cognition?

3. Explain the importance of the question, "Who's doing the talking?" in Bakhtin's theory.

Objective 5. **Describe apprenticeship as a way of learning, and discuss its classroom implications.**

A. *Apprenticeship as It Relates to Social Interaction Theories*

1. Briefly define apprenticeship.

2. Explain why social interaction theorists consider apprenticeship superior to other instructional approaches.

3. Define guided participation and explain why it is an important part of apprenticeship.

4. Define each of the following components of apprenticeship in the classroom.

modeling:

coaching:

scaffolding:

articulation:

reflection:

exploration:

B. *Situated Cognition*

1. Define situated cognition.

2. Why is it important for school tasks to be more like real-world tasks?

Objective 6.   **Describe Reciprocal Teaching.**
    A.   *Features of Reciprocal Teaching*

1. List the strategies taught in Reciprocal Teaching

2. Outline the processes of reciprocal teaching

Objective 7. **Describe studies that support and refute the claims of social interaction theorists.**

In the columns below, list findings of studies from throughout the chapter that support or refute the claims made by social interaction theorists. Be sure to note the specific claim or concept to which the research relates.

Supporting Research	Refuting Research

Objective 8.　　　　**Compare and contrast Initiate-Respond-Evaluate cycles with true academic dialogues, including the impact of these instructional conversations on students.**

　　A.　　*Initiate-Respond-Evaluate Cycles*

1. Briefly explain each component of the cycle and give an example.

Initiate

Respond

Evaluate

　　B.　　*Naturalistic Classroom Conversations*

1. List 4 structural differences between naturalistic academic conversations and other types of natural conversations.

a)

b)

c)

d)

2. Explain distributed cognition, which occurs in naturalistic classroom conversations.

C.   *Strengths and Limitations/Challenges of Types of Classroom Interaction*
Complete the following table. Note when a potential strength or limitation has been either supported or refuted by research mentioned in the chapter.

	Initiate-Respond-Evaluate	Classroom Conversations
Strengths		
Limitations and Challenges		

Objective 9. **Discuss factors that promote and inhibit effective academic discussions.**

In the table below, list instructional factors that can support or weaken academic discussions.

To Promote Academic Conversations:	To Inhibit Academic Conversations:

PRACTICE TESTS

[See answer key at the end of the chapter for correct responses.]

Multiple Choice

Circle the letter of the best response to each question.

1. Which of the following is true of infants (i.e., stage one) according to Vygotsky's theory?
   a) They have not developed a form of speech.
   b) Their thoughts are verbal.
   c) They use noises to communicate ideas.
   d) Their thoughts are completely separated from their speech.

2. According to Vygotsky's theory, which of the following attributes distinguishes a stage-three preschooler from a stage-two two-year-old?
   a) Thought and speech merge for the first time.
   b) The child begins verbally communicating with others.
   c) The child uses speech to direct her own thinking.
   d) The child starts naming objects, a sign that the child is developing conceptual speech.

3. Which of the following statements represents Vygotsky's fourth and final stage of thought/speech development?
   a) People "think out loud" to manage tasks.
   b) Thought and speech overlap completely.
   c) People silently talk themselves through tasks.
   d) People no longer need speech to think through a task.

4. Which of the following statements represents a difference between inner speech and outer speech?
   a) If a problem is familiar, an adult will use inner speech to work through it.
   b) Actions are more prominent in outer speech.
   c) Outer speech is in a form that is easy to understand, whereas inner speech is fragmentary and abbreviated.
   d) After inner speech is developed, the child learns to communicate it as outward speech.

5. Which of the following forms of speech does Vygotsky consider the most developmentally advanced?
   a) Outer speech
   b) Dialogue
   c) Egocentric speech
   d) Inner speech

6. According to Vygotsky's theory, cognition develops twice. Which of the following statements best describes this process?
   a) First the child thinks through a task in interaction with someone, then the child internalizes that thinking.
   b) First the child learns how to do the task by watching someone else do it, then the child learns how to do the task alone.
   c) First the child gets lots of praise when doing the task, then the child starts doing the task without praise.
   d) First the child gains declarative knowledge about how a task is done; procedural knowledge, or the ability to do the task, develops later.

7. Which of the following best represents scaffolding a task in a child's zone of proximal development?
   a) Zola does not know how to tie her shoes, so her mother does it for her.
   b) Cary has been responsible for completing his homework for several years, but his father still reminds him to do it anyway.
   c) Julie does not have the coordination to play the piano, so her aunt holds her hands and guides her fingers until they are on the correct keys.
   d) Joshua needs a little help drawing shapes; his teacher, who wants him to be independent, praises him when he does it correctly.

8. Which of the following is a feature of scaffolded instruction?
   a) The teacher models how to do the task and the learner just observes until he or she figures out how to do the task alone.
   b) The teacher provides constant support.
   c) The teacher allows the child to fail once in awhile.
   d) The teacher gives only as much prompting as the learner needs.

9. Which of the following statements is **not** consistent with Bakhtin's theory?
   a) What other people say shapes our ideas.
   b) When we speak, what we say represents internalized ideas of people with whom we interact.
   c) One's most predominant cultural group will come through in one's ideas and words regardless of the situation.
   d) The only "voices" that affect you will be those of individuals you directly interact with.

10. The idea that school environments should be restructured to include more real-world problem-solving is best represented by which of the following terms?
    a) Reciprocal teaching
    b) Distributed cogntition
    c) Situated cognition
    d) Guided participation

11. Which of the following components of apprenticeship requires the apprentice to explain how he or she is solving a problem?
    a) Modeling
    b) Articulation
    c) Reflection
    d) Exploration

12. Which of the following components of apprenticeship requires the apprentice to compare his or her work with the work of others?
    a) Coaching
    b) Articulation
    c) Reflection
    d) Exploration

13. Which of the following components of apprenticeship encourages the apprentice to develop his or her own style, different than the master's?
    a) Scaffolding
    b) Articulation
    c) Reflection
    d) Exploration

14. Which of the following is **not** a feature of apprenticeship?
    a) The master's role is to bridge gap between known and unknown, mainly by explaining how to do the task.
    b) The master increases expectations as apprentice gains capability.
    c) The apprentice's role is to observe and practice with coaching.
    d) It is important for the apprentice to learn to talk like the master.

15. Reciprocal Teaching is highlighted as being consistent with which of the following theorists?
    a) Bakhtin
    b) Piaget
    c) Palincsar
    d) Vygotsky

16. Which of the following features distinguishes Reciprocal Teaching from apprenticeship?
    a) The teacher models how to use strategies.
    b) A student leader guides the group's strategies use.
    c) The teacher provides support as it is needed.
    d) The teacher provides structured experiences for students to practice.

17. According the the chapter, which of the following is true of Reciprocal Teaching?
    a) It works better when the teacher explicitly teaches the strategies first.
    b) The teacher should remain an active participant in the group.
    c) Students tend to ask high-level questions.
    d) Studies showed that students monitored how well they understood what they were reading.

18. Which of the following claims has been supported by research?
    a) Adult-child scaffolding is universal
    b) Scaffolding is superior to other forms of instruction
    c) Adults excert less control as children become more capable.
    d) Children who get scaffolded instruction from a parent learn faster than children whose parents are less attentive to the child's competence and specific difficulties.

19. Which of the following is true of Initiate-Respond-Evaluate cycles of classroom interaction?
    a) Students need lots of guidance to interact this way.
    b) They help teachers cover key information efficiently.
    c) Discussion is often at a low cognitive level.
    d) Students control the interactions.

20. Which of the following distinguishes naturalistic classroom conversations from other types of conversation?
    a) Few people are involved.
    b) They involve close relationships.
    c) One participant has most of the control.
    d) There are more choices of topics.

21. Which of the following best represents the teacher's role in naturalistic classroom conversations?
    a) To initiate the conversation topic and ask questions.
    b) To lead the group to a specific answer.
    c) To evaluate whether students understand the topic.
    d) To make sure certain points are covered.

Matching

Match the letters of the description on the right with the corresponding numbered terms on the left. Use each description only once. Some descriptions may be left over.

_____ 1. Apprenticeship
_____ 2. Distributed Cognition
_____ 3. Initiate-Respond-Evaluate
_____ 4. Inner speech
_____ 5. Multivoicedness
_____ 6. Reading Recovery
_____ 7. Reciprocal Teaching
_____ 8. Scaffolding
_____ 9. Situated Cognition
_____ 10. Social language
_____ 11. Zone of Proximal Development

a) Involves guided participation from an expert
b) Speaking differently in different situations
c) Typical classroom interaction
d) Learning is often tied to the environment in which it is learned
e) Students take turns quizzing each other about a set of chemistry terms
f) A form of language that guides thinking
g) Form of language used in groups
h) Ideas emerge from the whole group's interactions rather than just from individuals
i) Doing a difficult task for a child who is unable to do it
j) Task child can do with help but not alone
k) Students in a group use a set of strategies to support reading comprehension
l) A form of learning in which each student is assigned part of the task and is responsible for a portion of the thinking
m) A form of speech used in a specific cultural situation
n) An instructional program in which an adult tutor gives hints and support so that a child will have success decoding a text.
o) Coaching that is faded when child becomes more independent

LEARNING STRATEGIES

Below are examples of strategies that can help you understand major chapter concepts. Use these examples to guide your own strategies.

Strategy Example #1

Think of a skill you learned or could learn through apprenticeship [for example, a sport, art, music, how to study...]. Explain the apprenticeship learning process, using this example.

Strategy Example #2

Choose a content area taught in school, and brainstorm real-life tasks that could be integrated into the classroom (situated cognition).

Strategy Example #3

Work with a classmate to understand the issues surrounding academic discussions. Assign each person a side "for" or "against" the use of academic conversations. Each person should argue for his or her assigned position, using information from the chapter.

ANSWER KEY

Strengthening What You Know

Objective/Item
3.

	VYGOTSKY	PIAGET
Time of their Research	[They were doing work around the same time.]	
Time of most influence	[70's, 80's and 90's] Explain why difference makes sense.	[Peaked in 60's and 70's]
Dialectical	[Development occurs due to interaction between individual and society.]	
Role of individual	[Individual is an active learner, and active exploration is key to development.]	
Dynamic Development	[Development is a dynamic process, full of upheaval &sudden changes.]	
ZPD / Disequilibrium	[Vygotsky's Zone of Proximal Development is similar to Piaget's idea of disequilibrium and cognitive conflict--both have the same principle of teaching just beyond the current competency level.]	
Role of social world	[Development of individual cannot be understood without understanding the social environment. Development is influenced by social institutions of the culture (schools, government) and tools of the culture (language, technology).]	[Social world influences are not central; child acts as an individual.]
Role of language **egocentric/ internalized speech**	[Egocentric speech represents an advance-- Communication with oneself learned after communication with others. Child uses egocentric speech to guide actions and work through challenging tasks. This speech does not disappear with maturity; rather, it becomes internalized as inner speech that has a major role in shaping thinking.]	[Speech is a "symptom" of ongoing mental activity. Egocentric speech represents a deficit--child unable to communicate meaning to others/ understand other's perspective. Egocentric speech disappears with maturity.]

Multiple Choice

Correct answers are in bold.
Comments related to other options indicate why that response is incorrect.

1.   **d)**

2.   **c)** [Others occur stage 2.]

3.   **c)**
       b) Not completely--see diagram.
       d) Actually, speech and language are even more integrated, rather than separated.

4.   **c)**
       a) For familiar problems, even inner speech is no longer necessary for adults.
       b) No--Actions are more prominent in inner speech.
       d) The process is reverse--the overt speech becomes covert.

5.   **d)**

6.   **a)**

7.   **c)**

8.   **d)**
       b) Support is gradually removed as the learner becomes more independent.

9.   c) The social context often determines how we will speak and which of our internalized voices comes through.
       d) The voices Bakhtin talks about often represent cultural groups as well as individuals; moreover, an individual's voice--as Bakhtin uses the term--can come through in writing, art, music, as well as direct conversations.

10.   **c)**

11.   **b)**

12.   **c)**

13.   **d)**

14.   **a)** Showing and guiding more than explaining
       d) This is a sign of cognitive change, and it brings the individual into the culture.

15.   **d)**

16.   **b)**

17.   **a)**

18.   **c)**

19.   **b)**

20.   **c)**

21.   **a)**

Matching

__a__ 1. Apprenticeship
__h__ 2. Distributed Cognition
__c__ 3. Initiate-Respond-Evaluate
__f__ 4. Inner speech
__b__ 5. Multivoicedness
__n__ 6. Reading Recovery
__k__ 7. Reciprocal Teaching
__o__ 8. Scaffolding
__d__ 9. Situated Cognition
__m__ 10. Social language
__j__ 11. Zone of Proximal
         Development

<u>Correct terms for responses left over</u>:
e) [none--distractor for reciprocal teaching]
g) [none--distractor for social language]
i) [none--distractor for scaffolding]
l) [none--distractor for distributed cognition]

CHAPTER 9

Social Influences in the Classroom

LEARNING OBJECTIVES

1. Describe key development changes in self-concept through the school years, and discuss ways to enhance self-concept and use it as a basis for learning information.

2. Describe the 8 life stages in Erikson's theory of identity development, and identify situations that represent the key conflict at each stage.

3. Explain how relationships with parents and peers can affect academic achievement.

4. Describe Piaget's theory of how moral judgment develops.

5. Describe Kohlberg's stages of moral reasoning and identify examples of reasoning at each stage.

6. Discuss criticisms of Kohlberg's theory and describe potential gender and cultural differences in moral reasoning.

7. Compare and contrast approaches to moral education, including ethics courses, Kohlberg's moral dilemmas, Kohlberg's just communities, and Lickona's integrative approach.

STRENGTHENING WHAT YOU KNOW

The purpose of this chapter is to examine key social influences on academic achievement and to explore moral development and moral education.

Objective 1. **Describe key development changes in self-concept through the school years, and discuss ways to enhance self-concept and use it as a basis for learning information.**

    A.   *Developmental Changes in Self-Concept*

1. Define self-concept.

2. In the table below, characterize the development of self-concept in the school years by noting key differences between younger and older students.

Younger Students	Older Students

B.   *Enhancing Learners' Self-concept in the Classroom*

1. List the chapter's "general strategies for enhancing children's self-concepts," and include a brief possible explanation of why each approach works.

a)

b)

c)

d)

e)

C.   *Self-Schemata as a Basis for Learning*

1. Define self-schema.

2. How can a person's self-schema help when learning information in school?

Objective 2.   **Describe the 8 life stages in Erikson's theory of identity development, and identify situations that represent the key conflict at each stage.**

A.   *Central Conflicts*

Describe the role of conflict in Erikson's theory.

B.   *Stages of Identity Development*

Complete the table below to summarize Erikson's stage theory of identity development.

Age span	Central Conflict	Give an example of an event that represents the central conflict. Explain the impact of positive & negative resolutions.	What parent/teacher actions can foster or impede positive resolution?

C. *Identity Statuses*

The four identity statuses differ with respect to the experience of crisis and commitment to an identity. In the matrix below place the statuses--moratorium, identity achieved, diffusion, and foreclosure--to illustrate their relationships. Include a description of each.

	Crises	No Crises
Commitments		
No Commitments		

D. *Ethnic Identity*

1. What kinds of impact does ethnicity have on identity development?

2. What can teachers do to encourage healthy identity development of minority students?

Objective 3. **Explain how relationships with parents and peers can affect academic achievement.**

    A.   *Parents*

1. Complete the following matrix to show the relationships of 4 kinds of parenting: authoritarian, authoritative, indulgent/permissive, neglecting. Include examples of behavior for each type of parent.

	Demanding	Undemanding
Responsive		
Unresponsive		

2. Strategy--How will you distinguish the terms authoritarian and authoritative?

B.   *Peers*
1. Describe developmental differences in friendships in the chart below.

Preschool Friendships	Middle Childhood/Adolescent Friendships

2. Describe Selman's levels of social competence in the chart below.
For each level, create an example of 2 children interacting.

LEVEL OF SOCIAL COMPETENCE	DESCRIPTION, TYPICAL BEHAVIORS, EXAMPLE
Impulsive	
Unilateral	
Reciprocal	
Collaborative	

C.   *Combined Influences of Parents and Peers, and the Role of Culture*
1. What explanation is given for cultural differences in parental impact on achievement?

2. Describe how peer relationships can moderate the impact of parenting style on achievement.

Objective 4. Describe Piaget's theory of how moral judgment develops.

A. *Piaget's Stages of Moral Development*

Compare and contrast Piaget's 2 stages of moral development in the chart below.

	View of Objective Consequences vs. Intentions	View of Rules
Heteronomous Morality		
Autonomous Morality		

B. *Processes of Moral Development*

1. According to Piaget, how does a child's moral reasoning move from heteronomous to autonomous?

2. Based on this view, what could a teacher or parent do to encourage a child's moral reasoning development? (Give a few suggestions)

Objective 5. **Describe Kohlberg's stages of moral reasoning and identify examples of reasoning at each stage.**

Complete the following table to organize information about Kohlberg's stages.

Stage	Name/Label	Description/Example
PRE-CONVENTIONAL		
1		
2		
CONVENTIONAL		
3		
4		
POST-CONVENTIONAL		
5		
6		

Objective 6. **Discuss criticisms of Kohlberg's theory and describe potential gender and cultural differences in moral reasoning.**

    A.   *Role of Gender in Reasoning*

1. Describe Gilligan's criticisms of Kohlberg's theory and Gilligan's view of a more appropriate theory.

2. What evidence supports, or does not support, Gilligan's criticisms?

Evidence supporting the criticism	Evidence not supporting the criticism

    B.   *Cultural Influences on Moral Reasoning*

1. Describe criticisms of the universality of Kohlberg's theory.

2. What evidence supports, or does not support, this criticism?

Evidence supporting the criticism	Evidence not supporting the criticism

Objective 7. **Compare and contrast approaches to moral education, including ethics courses, Kohlberg's moral dilemmas, Kohlberg's just communities, and Lickona's integrative approach.**

1. Complete the following table comparing and contrasting the approaches.

APPROACH	BRIEF DESCRIPTION OF INSTRUCTIONAL APPROACH Note Distinguishing Features	IMPACT ON MORAL REASONING; NOTE STRENGTHS/WEAKNESSES
Ethics Courses		
Social studies topics		
Plus-one Approach		
Just Community		
Lickona's Integrative Approach		

2. How can teachers encourage thoughtfulness in their classrooms?

3. What obstacles make it hard to implement the thoughtful education described above?

4. If you plan to teach, what will you do to overcome these obstacles?

PRACTICE TESTS

[See answer key at the end of the chapter for correct responses.]

Multiple Choice

Circle the letter of the best response to each question.

1.   Which of the following is true of self concept?
     a) It is stable over time.
     b) It is an organized representation of ideas.
     c) It is constant from one situation to another.
     d) Teachers have little impact on a student's self-concept.

2.   Which of the following terms probably best describes the self-concept of an older elementary student?
     a) Generalized
     b) Effort-based
     c) Abstract
     d) Over-inflated

3.   Which of the following examples probably best describes the self-concept of a younger elementary student?
     a) Kurt tells his teacher, "I'm a very good reader, but I'm not as good at science."
     b) LaShawn compares herself with her friend Sari, noticing that Sari spends as much time on homework but does not seem to be quite as smart.
     c) Joshua describes himself as shy, polite, and kind.
     d) Ana tells her father, "I know I am very smart because I have a lot of friends at school."

4.   Which of the following terms best describes Kurt's self-concept in example "a" above (item 3)?
     a) concrete
     b) abstract
     c) generalized
     d) differentiated

5. Which of the following is true of the development of self-concept?
   a) Compared to young children, older children are more likely to describe themselves by citing specific behaviors.
   b) Older children are more self-confident in school because they are more cognitively advanced than younger children.
   c) Older children, but not younger children, understand that easy tasks require less ability.
   d) Older children view ability as a changeable trait.

6. To a junior high school student, which of the following situations would most likely be considered an indicator of high ability?
   a) The student gets one hundred percent on a fourth-grade spelling test.
   b) The teacher says, "Good work!" when the student does well on the fourth-grade spelling test.
   c) The teacher gets mad when the student does poorly on a fourth-grade spelling test.
   d) The teacher says, "I'm sorry you didn't do well" when the student does poorly on a fourth-grade spelling test."

7. Which of the following helps explain developmental changes in self-concept?
   a) Students don't try as hard in school as they get older and other issues are on their mind.
   b) Young students are more affected by teachers reactions to their success and failure on simple assignments.
   c) Teachers of older students are more likely to reward effort than are teachers of younger students.
   d) Younger students think they are smart when they work hard, whereas older children see this as two separate issues.

8. Which of the following approaches is **not** considered effective in enhancing students' self-concept?
   a) Show students strategies that will help them do difficult work.
   b) Help students a little to make sure they experience success on meaningful task.
   c) Value students suggestions and opinions.
   d) Let smarter students know that they are at the top of the class.

9. Which of the following class activities best encourages self-reference as defined in the chapter?
   a) Have students list 10 of their own positive characteristics and then guess what other students selected as their personal positive traits.
   b) Have students evaluate their personal progress by charting changes in their own grades over time.
   c) Have students make a collage to describe themselves.
   d) Have students role-play significant figures in your content area in a mock talk-show interview.

10. Which of the following is true of Erikson's theory of identity development?
    a) Development is based on conflict.
    b) Each stage has many potential outcomes.
    c) Each stage is independent of the other stages.
    d) Once a child has completed a stage, it is no longer an issue for him/her.

11. Which of the Erikson's stages is most closely tied to learning academic skills?
    a) Initiative vs. Guilt
    b) Autonomy vs. Shame and Doubt
    c) Industry vs. Inferiority
    d) Identity vs. Identity confusion

12. In which of the following stages would a child most strongly develop attitudes and preferences regarding sexuality?
    a) Generativity vs. Stagnation
    b) Intimacy vs. Isolation
    c) Identity vs. Identity confusion
    d) Trust vs. Mistrust

13. Monica is constantly trying new clothing and hairstyles, yet rarely seems satisfied with how she looks. She likes to fill out self-questionnaires in magazines like, "Finding Your Most Compatible Date," and she spends lots of time on the phone with friends, comparing opinions and ideas. What is Monica's most likely identity status?
    a) moratorium
    b) identity achieved
    c) diffusion
    d) foreclosure

14. Which of the following is true of ethnic identity?
    a) Ethnic minority adolescents have a higher rate of identity moratorium than do adolescents of the majority culture.
    b) Experiencing prejudice tends to weaken a person's ethnic identity.
    c) Being an ethnic minority tends to broaden adolescents' identity possibilities, as they have choices from both the minority and the majority culture.
    d) Identity development for ethnic minorities may be hampered because minorities are often excluded or negatively portrayed in school materials.

15. Which type of parenting helps children internalize behavioral standards?
    a) authoritarian
    b) authoritative
    c) indulgent/permissive
    d) neglecting

16. Which of the following is true of peer relationships?
    a) Expectations of friendships stay about the same over the school years.
    b) Peers can negatively, but not positively, change the impact parents have on academic achievement.
    c) Peers can reinforce and punish one another.
    d) As children get older, they find it more difficult to integrate several friends' perspectives.

17. Which of the following is a focus of preschool friendships?
    a) Loyalty
    b) Mutual understanding
    c) Common activities
    d) Intimacy

Items 18-21 below are based on this situation: Jamal is friends with Marilyn and Juan, who are dating. One day Marilyn tells Jamal that before she dated Juan, she slept with someone who recently tested HIV positive. Marilyn does not plan to tell Juan, and she asks Jamal not to say anything to him about it. For each of the items below, indicate which of Kohlberg's moral stages is represented by Jamal's decision.

18. Jamal tells Juan because he thinks it is his responsibility to let Juan know.
    a) 1
    b) 2
    c) 3
    d) 4
    e) 5
    f) 6

19. Jamal decides that Juan's life is in danger and that this is more important than keeping a promise or maintaining a friendship. He tells Juan.
    a) 1
    b) 2
    c) 3
    d) 4
    e) 5
    f) 6

20. Jamal decides not to tell Juan; he refuses to violate his principle of integrity, which includes not breaking confidences.
    a) 1
    b) 2
    c) 3
    d) 4
    e) 5
    f) 6

21. Jamal decides not to tell Marilyn's secret because if he did, she would be angry and might end their friendship.
    a) 1
    b) 2
    c) 3
    d) 4
    e) 5
    f) 6

22. Which of the following is **not** one of the criticisms of Kohlberg's work?
   a) He got his ideas from Piaget without doing much of his own work.
   b) His views do not apply to all cultures.
   c) His views don't account for a focus on interpersonal relationships.
   d) His theory is based on research with boys.

23. Which of the following researchers argued that Kohlberg's theory rates females as having lower moral reasoning than males?
   a) Piaget
   b) Gilligan
   c) Bandura
   d) Erikson

24. Which of the following instructional approaches has been least successful in fostering the development of moral reasoning?
   a) Plus-one approach
   b) Philosophy courses in ethics
   c) Just Community
   d) Lickona's Integrative approach

25. Which of the following is a drawback of Kohlberg's programs of moral instruction?
   a) They don't improve students' moral reasoning.
   b) They demand strong social and communication skills of students.
   c) They attempt to directly teach students about moral behavior.
   d) Students tend to set easy rules and consequences, resulting in many discipline problems.

26. Which of the following is **not** a component of Lickona's integrative approach to moral education?
   a) Students freely express their views and teachers do not judge students' perspectives.
   b) Teachers include literature/storytelling that emphasizes moral themes.
   c) Teachers ask students to behave in a more ethical manner.
   d) Teachers use punishment and reinforcement to discourage immoral behavior and to encourage moral behavior.

Completion

Fill in each blank with the best fitting term from the chapter. Terms are used only once.

1. The least healthy identity status is probably _____.

2. Harold he decided without question to attend his father's alma mater and to major in literature as everyone thinks he should, and he is perfectly happy with his decision. His identity is in _____.

3. The most healthy progression of identity statuses is to first experience _____.

4. Parents who use coercion to control children's behavior are _____.

5. Understanding one's membership in a cultural group is _____ identity.

6. According to Piaget, the central process in moral reasoning development is _____.

7. Through imitation, children can learn to make more mature moral judgments, consistent with _____ theory.

8. In Kohlberg's measure of student's moral reasoning on dilemmas, the most important indicator is students' _____.

9. Maintaining social order is the primary concern in Kohlberg's _____ stage of moral judgments.

10. Kohlberg's _____ stage is based on self-interest.

11. Students are involved in setting rules and consequences in the _____ approach to moral education.

12. Discussions expose students to reasoning just beyond their current thinking in _____ education.

13. Behavioral and cognitive interventions are included in _____ moral education.

Matching

Match the letters of the description on the right with the corresponding numbered terms on the left. Use each description only once. Some descriptions may be left over.

_____ 1. Autonomous
_____ 2. Collaborative
_____ 3. Dualistic
_____ 4. Heteronomous
_____ 5. Impulsive
_____ 6. Reciprocal
_____ 7. Relativistic
_____ 8. Self-concept
_____ 9. Self-reference
_____ 10. Self-schema
_____ 11. Unilateral

a) Don't understand that others interpret the same behavior differently; can't distinguish action from feeling

b) Believing that there is only one right viewpoint

c) Belief that one has what it takes to do well on academic tasks

d) Coordinate perspectives through compromise to meet mutual goals

e) Information is easier to remember when you connect it to your views of yourself.

f) A motivational construct in which the person tries to reduce the difference between their current state and what they know they can become

g) Resolve conflicts by either controlling or giving in

h) Organizes our interpretations of our experiences and guides behavior

i) Stage of morality focusing on consequences and unbreakable rules

j) Believing the most moral choice is based on the context

k) Make deals and trades instead of integrating perspectives

l) A representation of theories, attitudes, and beliefs about oneself

m) One's views about whether success or failure depends on controllable or uncontrollable causes.

n) Moral stage in which rules can be changed and intentions are more important than outcomes

LEARNING STRATEGIES

Below are examples of strategies that can help you understand major chapter concepts. Use these examples to guide your own strategies.

Strategy Example #1

Choose one of the chapter's suggested activities teachers can use to help students develop a positive self-concept. Based on the chapter's discussion of self-concept development, role-play how a young child would most likely respond to this task (It might be fun to actually try to do the activity, thinking back on yourself as a young child). Next, role-play an older child responding to the same activity. This role-play may bring alive some of the developmental differences in self-concept.

Strategy Example #2

Choose a challenging concept or set of concepts in the chapter and decide how you can use the self-reference effect to remember and understand the information.

Strategy Example #3

Evaluate one of the courses you are taking according to the "criteria for promoting thoughtfulness." You can also compare two or more courses on these criteria.

ANSWER KEY

Multiple Choice

Correct answers are in bold.
Comments related to other options indicate why that response is incorrect.

1. **b)**
    a) It is dynamic and develops throughout the lifespan.
    c) It is dynamic and varies with the situation.

2. **c)**

3. **d)** does not differentiate academic ability from social skills
    a) differentiated self-concept
    b) distinguishes effort from ability
    c) abstract

4. **d)**

5. **c)**
    b) Older students are less confident.

6. **c)** Considered a sign that the student could do better
    d) Sympathy for failure indicates teacher thinks student can't do better.

7. **d)** [Other statements are False.]

8. **d)** Rather, encourage students to use their previous performance as their standard of success.

9. **d)**
    a) enhances self-concept, but does not tap the self-reference effect
    b) enhances self-concept, but does not tap the self-reference effect
    c) enhances self-concept, but does not tap the self-reference effect

10. **a)**
    b) Two major outcomes
    d) Sometimes negative outcomes are resolved later in development.

11. **c)**

12. **c)**

13. **a)**

14. **d)**
   a) No, ethnic students have a higher rate of identity foreclosure.
   b) Sharing experiences of prejudice fosters ethnic identity. Group members identify with membership in the minority group when they see others experiencing the same prejudices. Often this motivates desire to understand one's culture.
   c) Cultural pressures often limit rather than expand a person's perceived options.

15. **b)**

16. **c)**
   b) Peers can negatively and positively modify the impact of parenting styles.
   a) Expectations of friendships stay about the same over the school years.

17. **c)**

18. **d)**

19. **f)**

20. **e)**

21. **a)** Jamal fears the punishment of Marilyn ending their friendship.

22. **a)** Kohlberg conducted research and significantly expanded on Piaget's views.

23. **b)**

24. **b)**
   a) moral dilemmas

25. **b)** These demands make the programs challenging to carry out.
   c) Ethics courses have this feature.
   d) Not true based on text.

26. **a)** Lickona rejects "values clarification" that suggests morals are based on opinion.

Completion

1. diffusion
2. foreclosure
3. moratorium
4. authoritarian
5. ethnic
6. equilibration or cognitive conflict
7. social learning
8. reasons
9. conventional
10. pre-conventional
11. just community
12. plus-one
13. integrative; Lickona's approach

Matching

__n__ 1. Autonomous
__d__ 2. Collaborative
__b__ 3. Dualistic
__i__ 4. Heteronomous
__a__ 5. Impulsive
__k__ 6. Reciprocal
__j__ 7. Relativistic
__l__ 8. Self-concept
__e__ 9. Self-reference
__h__ 10 Self-schema
__g__ 11. Unilateral

Correct terms for responses left over:
c) self-efficacy
f) possible selves
m) an aspect of attributions

CHAPTER 10

Schooling Practices

LEARNING OBJECTIVES

1. Describe the characteristics of schools that promote good thinking.

2. Explore the role of the following school and classroom practices in student achievement: questioning, expectancies, peer tutoring, ability grouping, homework, extracurricular activities, and class size.

3. Compare and contrast the following instructional methods: direct instruction, direct explanation/teacher modeling, reciprocal teaching, discovery learning, cooperative learning, mastery learning.

4. Explore four components of effective classroom management: room arrangement; rules and procedures; communication; and intervention for problems.

STRENGTHENING WHAT YOU KNOW

The purpose of this chapter is to examine the strengths and weaknesses of a variety of classroom practices and instructional methods. The chapter also introduces principles of effective classroom management.

Objective 1.   **Describe the characteristics of schools that promote good thinking.**

    A.   *School Characteristics Associated with High Achievement*

1. Combine the findings of the two studies on schools that promote achievement to make a master list of characteristics. Next to each characteristic, write an explanation--why does it make sense that this is important?

Characteristics of Schools with High Achievement	Your Reasons for Why This Characteristic Is Important

B.   *Engaging Educational Environment*

1. What does "engagement in learning" mean, and why is it important?

2. In the table below, list the characteristics of engaging environments and add your own explanation for why each is important.

Characteristics of Engaging Classrooms	Your Reasons Why This Is Important

Objective 2.   **Explore the role of the following school and classroom practices in student achievement: questioning, expectancies, peer tutoring, ability grouping, homework, extracurricular activities, and class size.**

    A.   *Classroom Questioning*

1. Compare and contrast higher and lower order questioning in the following table:

	Higher Order Questioning	Lower Order Questioning
Definition		
Examples		
Results with high-ability students		
Results with low-ability students		
Results with mixed abilities		
Impact on discussion		

2. Describe wait time and its role in achievement.

3. How long does the teacher need to wait for a student to answer a question? How could the child's culture affect the answer to the previous question?

B.   *Teacher/Student Expectancies*

1. Explain the role of teacher expectancy on student achievement.

2. How consistent are the effects of teacher expectancies on student achievement?

3. How consistent are the effects of teacher expectancies on teacher behavior toward students?

4. Contrast teacher's behavior toward students they perceive as high-ability vs. low ability:

Teacher behavior	High-ability students	Low-ability students
Demands on student		
Wait time		
Responses to student questions		
Criticism		
Praise		
Reinforcements Contingencies		
Friendliness		
Calling on students		
Seating students		
Grading strictness		

5. How would you summarize the findings outlined in the above table?

6. What is meant by self-fulfilling prophecy, and what are its implications for teachers?

7. How can students' expectancies affect a teacher's performance?

     C.   *Peer Tutoring*

1. What affect does peer tutoring have on the student being tutored?

2. What affect does peer tutoring have on the student doing the tutoring?

3. At what age levels is peer tutoring effective?

D. *Ability Grouping/Tracking*

1. What is the impact of ability grouping or tracking on achievement?

2. In the table below contrast instruction in high-ability and low-ability groups:

High-ability groups	Low-ability groups

3. Explain U. S. Public Law 94-142.

4. What does research suggest about the effectiveness of mainstreaming compared to resource room instruction? What makes the less effective approach worse for students?

E.   *Homework*

1. Research suggests that homework promotes learning, but that its benefits are greater for some purposes than for others. Fill in the table to contrast high/low impact areas.

Homework Is Most Useful For:	Homework Is Less Useful For:

2. A father asks you whether he should help his children with their homework. What advice would you give him and why?

F.   *Extracurricular Activities*

1. What does research suggest about the affect of extra-curricular activities on school achievement?

2. What are some potential problems associated with students who have after-school employment?

G.   *Class size*

1. What is the main finding about the impact of class size on student achievement?

2. What is the ideal class size? Give the two views presented in the chapter.

Objective 3.   **Compare and contrast the following instructional methods: direct instruction, direct explanation/teacher modeling, reciprocal teaching, discovery learning, cooperative learning, mastery learning.**

A.   *Comparison of Instructional Features*

1. Complete the following table comparing the approaches on several categories.

	Amount of Teacher-directedness	Amount/ Type of Structure	Teacher's Role	Students' Role	Key/ Distinguishing Features
Mastery Learning (ML)					
Direct Instruction (DI)					
Direct Explanation (DE)					
Reciprocal Teaching (RT)					
Cooperative Learning (CL)					
Discovery Learning (DL)					

2. In the line before each instructional feature, put the initials of any of the instructional approaches including that feature.

_____ Focuses on efficient use of time
_____ Teaches strategies for learning
_____ Teacher quickly reduces control of instruction.
_____ Involves drill of information
_____ Involves exploring, manipulating, experimenting
_____ Emphasizes student's thought processes
_____ Student-directed
_____ Goal is high accuracy in responding, sometimes called "overlearning."
_____ Can be inefficient
_____ Includes mental modeling
_____ Emphasizes clearly specifying learning goals to students
_____ Based on cultivating natural curiosity
_____ Uses Vygotskian principles of scaffolding
_____ Uses Piagetian principles of equilibration
_____ Includes individual testing, and often group rewards
_____ Includes frequent testing
_____ Student has role of group leader.
_____ Compares students to own past performance
_____ Includes responsive elaboration
_____ Increases risk of incorrect learning
_____ Teams compete with each other.
_____ Presents content in a sequence, structured fashion
_____ Teacher give support as it is needed.
_____ Students work at their own pace.
_____ Limits teacher's role; teacher does not present information.
_____ Uses small steps at a rapid pace
_____ All students can get an A, even if it takes some longer than others.
_____ Includes immediate feedback
_____ Gives students tasks they can handle with teacher's help
_____ Based on task-oriented, noncompetitive approach
_____ Emphasizes review of previous material
_____ Teacher maintains control of instructional flow.
_____ Includes teacher presentations/explanations
_____ Teacher asks questions and evaluates students' responses.

3. In direct instruction, what are four features instructors should include in their feedback to students?

a)

b)

c)

d)

4. What are the four essential characteristics of cooperative learning?

a)

b)

c)

d)

Objective 4. **Explore four components of effective classroom management: room arrangement; rules and procedures; communication; and intervention for problems.**

A. *Arranging the Classroom*

1. What are three key goals behind classroom arrangement?

a)

b)

c)

2. Briefly list the guidelines for classroom arrangement.

a)

b)

c)

d)

B. *Establishing Rules and Procedures*

1. Explain the difference between rules and procedures.

2. How can teachers communicate rules and procedures?

3. What role can students have in forming rules and procedures?

C.   *Communicating Effectively*

1. List key features to contrast constructive assertiveness with hostility and timidity.

Constructively Assertive	Hostile	Timid

2. Define empathetic listening and give an example.

3. Define "I-message" and give an example.

D. *Intervening when Problems Arise*

1. What is "withitness," and why is it important?

2. What approaches are most recommended for dealing with classroom problems? Note when the teacher would need to be especially sensitive to differences in student responses to the approach.
a)

b)

c)

d)

e)

f)

g)

3. List higher-level interventions teachers can try when the above approaches fail:
a)

b)

c)

d)

4. What role of punishment was suggested in the chapter?

5. What new features or principles of punishment did you learn in this chapter?

PRACTICE TESTS

[See answer key at the end of the chapter for correct responses.]

Multiple Choice

Circle the letter of the best response to each question.

1.   Which of the following is one of the key characteristics of schools where achievement is high?
     a) The environment is strict.
     b) They emphasize creativity and fun.
     c) The administrators have teachers take over decision-making.
     d) They set high demands on students.

2.   Which of the following classroom practices is most beneficial for student achievement?
     a) Have about 25 students in the class.
     b) Divide the class into high, average, and low reading groups.
     c) Keep the pace by moving on if a student does not answer a question.
     d) Ask questions students can't answer by just remembering information.

3.   Which of the following is **not** a key characteristic of an engaged learner?
     a) They care about their grades and success in school.
     b) They are competitive with other students.
     c) They are interested in academic content.
     d) They are committed to understanding the information in school.

4.   Which of the following is an important way to promote engagement with learning?
     a) Try to avoid tasks that may lead to student failure.
     b) Stay consistent in methods of teaching.
     c) Talk about what students want to learn.
     d) Allow students time to memorize what the teacher says.

5.   Which of the following after-school activities is most likely to have a negative impact on student achievement?
     a) Working the evening shift to earn date money.
     b) Attending late-night rehearsals to play the lead role in the class musical.
     c) Tutoring a classmate in social studies.
     d) Going to football practice and traveling to games.

6. Which of the following is an example of a higher-order question?
    a) What is the definition of a higher-order question?
    b) What kinds of students benefit most from higher-order questions?
    c) Which of the following is an example of a higher order question?
    d) Name four aspects of questioning that can affect achievement.

7. Mr. Anachi's students are answering questions with brief responses. Which action is probably the most appropriate for encouraging longer responses?
    a) Give students more wait time.
    b) Ask higher-level questions.
    c) Tell students to use complete sentences.
    d) Refer the questions to students from a more verbal cultural background.

8. According to the chapter, which of the following teacher behaviors is most likely directed at a student the teacher perceives as low-ability?
    a) Giving the student extra time to respond.
    b) Giving a more detailed explanation when the student has a question.
    c) Giving reinforcement even though the student's response was incorrect.
    d) Calling on the student often to check his or her understanding.

9. Which of the following statements is **not** true of peer tutoring?
    a) A child is more likely to listen to a brother or sister than to someone from outside the family.
    b) It helps the person being tutored, but it may hold the tutor back slightly.
    c) It is effective at the college level.
    d) It is effective at the elementary level.

10. What do low-track classes have more of than high-track classes?
    a) Homework
    b) Off-task time
    c) Exposure to learning
    d) Instruction

11. Which of the following instructional settings does the chapter most support for special education students?
    a) Private special education schools
    b) Separate classes in public schools
    c) Resource room instruction for part of the day and regular classroom instruction for the rest of the day
    d) Regular classroom instruction

12. Which of the following instructional approaches is most likely to include Initiate-Respond-Evaluate cycles of teacher/student communication?
    a) Mastery Learning
    b) Direct Instruction
    c) Direct Explanation
    d) Reciprocal Teaching

13. Which of the following represents a difference between direct instruction and direct explanation?
    a) Using students' time well
    b) Focusing on student processing
    c) Giving immediate feedback
    d) Specifying instructional goals

14. Which of the following represents a difference between direct instruction and reciprocal teaching?
    a) Scaffolding
    b) Teaching cognitive strategies
    c) Modeling and explaining thinking
    d) Students taking control quickly

15. Which of the following is **not** true of cooperative learning?
    a) Groups compete with each other
    b) Individual scores are included in a group grade
    c) Each person's score is based on changes from past performance
    d) The teacher works with groups as a tutor

16. Which of the following is true of mastery learning?
    a) Students work at their own pace.
    b) About half of the students can get an A.
    c) Students compete for the title of Master.
    d) Students can work on the units in any order they want.

17. Which of the following is least important in having a well-managed classroom?
    a) Being prepared with a variety of possible punishments for misbehavior.
    b) Leaving plenty of room between students' desks.
    c) Having class rules.
    d) Showing empathy when listening to students.

18. Which of the following is a principle of punishment?
   a) Punish students whenever they disobey rules.
   b) Let students know what kind of behavior will earn praise.
   c) Punishment increases the likelihood of desired behaviors.
   d) If the student is not visibly upset by the punishment, it probably wasn't strong enough.

19. A student who is normally well-behaved misbehaves in class. Which of the following is probably the most appropriate teacher reaction?
   a) Call the student's parents and tell them about the misbehavior.
   b) Give the student time out.
   c) Write a contract in which the student agrees not to repeat such behavior.
   d) Specifically praise other students for acceptable behavior.

20. Which of the following is an "I message"?
   a) I hear what you are saying. You feel angry that I did not choose you to feed the fish.
   b) I will give you two choices: You can put away the comic book and I will let you participate in the discussion, or you can continue reading the comic book and I will give you a red mark for the day.
   c) Drawing pictures in the class dictionary damages the book and distracts others who use it, which makes me angry.
   d) I would feel much better if you took your feet off Monica's desk.

21. Which of the following is **not** a main goal of classroom management?
   a) Maintain order
   b) Facilitate transitions from one lesson to another
   c) Make sure students raise their hands before speaking
   d) Figure out what students are doing

22. Which of the following would be useful in setting classroom rules/procedures?
   a) Have rules covering as many specific situations as you can think of.
   b) Tell students the rules once at the beginning of the year and let them know they will be responsible for remembering them.
   c) Make sure that procedures do not allow students to talk between lessons.
   d) Have students discuss why the rules are important.

Completion

Fill in each blank with the best fitting term from the chapter. Terms are used only once.

1. A common practice that does **not** promote academic achievement is grouping students by _____ .

2. Teachers can increase student achievement by giving students a chance to think before answering a question, also called _____.

3. U. S. Public Law 94-142 requires that special education students be place in _____ environments.

4. _____ works best with diverse groups, but not groups that include the full range of ability.

5. In a conflict, you can show the other person that you understand and respect his or her feelings by using _____ listening.

Matching

Match the letters of the description on the right with the corresponding numbered terms on the left. Use each description only once. Some descriptions may be left over.

_____ 1. Constructive assertiveness
_____ 2. Higher order questioning
_____ 3. Lower order questioning
_____ 4. Mental modeling
_____ 5. Responsive elaboration
_____ 6. Self-fulfilling prophecy
_____ 7. Student engagement

a) Thinking a student is smart or dumb affects behavior toward the student and may actually affect the student's achievement.

b) Requires student to remember information

c) Supporting one's own rights without attacking others

d) Enthusiasm, interest and commitment to learning

e) Showing students a replica of the human brain

f) Showing students how to do a task by thinking aloud while doing a similar task

g) Requires student to manipulate information

h) Giving a lecture about how the mind works

i) Asking a question that most students are unable to answer

j) A state occurring just before one of your pupils is about to get married

k) Giving feedback that is tied to students' specific needs and problems

LEARNING STRATEGIES

Below are examples of strategies that can help you understand major chapter concepts. Use these examples to guide your own strategies.

Strategy Example #1
Using some simple academic material or a concept you have learned in college, develop a sample lesson following the direct instruction approach. Label each of the direct instruction functions.

Strategy Example #2
In one high school students made a humorous video about the school rules. The video showed some students following rules and some students breaking rules. Do you think this was a good idea? Why or why not?

Strategy Example #3
Think of a current or recent conflict you have had with another person. Create specific examples of how you could use constructive assertiveness, empathetic listening, and/or eye messages to help resolve the conflict.
OR
Create a hypothetical conflict situation and role-play, with a partner, timid, hostile, and constructively assertive ways of communicating.

ANSWER KEY

Strengthening What You Know

Objective/Item
3. A. 2.

_____ Focuses on efficient use of time [DI, DE]
_____ Teaches strategies for learning [DE, RT]
_____ Teacher quickly reduces control of instruction [RT]
_____ Involves drill of information [DI, often ML]
_____ Involves exploring, manipulating, experimenting [DL]
_____ Emphasizes student's thought processes [DE, RT]
_____ Student-directed [DL; CL somewhat, but teacher maintains role as tutor;
    ML only in sense that student controls the pace]
_____ Goal is high accuracy in responding, sometimes called "overlearning" [ML, DI]
_____ Can be inefficient [DL]
_____ Includes mental modeling [DE, RT]
_____ Emphasizes clearly specifying learning goals to students. [DI, DE]
_____ Based on cultivating natural curiosity [DL]
_____ Uses Vygotskian principles of scaffolding [DE, RT]
_____ Uses Piagetian principles of equilibration [DL]
_____ Includes individual testing, and often group rewards [CL]
_____ Includes frequent testing [ML, DI]
_____ Student has role of group leader [RT]
_____ Compares students to own past performance [CL]
_____ Includes responsive elaboration [DE]
_____ Increases risk of incorrect learning [DL]
_____ Teams compete with each other [none]
_____ Presents content in a sequence, structured fashion [ML, DI]
_____ Teacher give support as it is needed [DE, RT]
_____ Students work at their own pace [ML]
_____ Limits teacher's role; teacher does not present information [DL]
_____ Uses small steps at a rapid pace [DI; not ML because student sets pace]
_____ All students can get an A, even if it takes some longer than others [ML]
_____ Includes immediate feedback [DI, DE]
_____ Gives students tasks they can handle with teacher's help [DI, DE]
_____ Based on task-oriented, noncompetitive approach [ML, possibly CL]
_____ Emphasizes review of previous material [DI]
_____ Teacher maintains control of instructional flow [DI, DE]
_____ Includes teacher presentations/explanations [DI, DE, RT, sometimes CL]
_____ Teacher asks questions and evaluates students' responses [DI]

Multiple Choice

Correct answers are in bold.
Comments related to other options indicate why that response is incorrect.

1. **d)**
   b) They emphasize academic skills.

2. **d)** Higher-order questions requiring students to reflect and manipulate information are better.
   a) Smaller classes (some say 15 is small enough, others say 1-3 students) are associated with achievement.

3. **b)**

4. **c)**

5. **a)**

6. **c)**

7. **b)**

8. **c)** Reinforcement is less likely to be contingent on correct responding for low-ability students.

9. **b)**

10. **b)**

11. **d)**

12. **b)**
    a) Most often independent work with less teacher-student interaction

13. **b)** direct explanation only

14. **d)**

15. **a)**

16. **a)**
   b) All students can get an A.
   d) Logical sequence; each one builds on the previous.

17. **a)** Punishment is a last resort.

18. **b)**
   a) Reserve punishment for epeated offenses, when reinforcing good behavior has not been enough.
   d) The punishment should fit the offense; punishment can backfire if it makes students too nervous or angry.

19. **d)**

20. **c)**
   a) empathetic listening
   b) emphasizes responsibility for self-control, but it's not an I-message

21. **c)** Although it may help maintain order, it is not the ultimate goal.

22. **d)**

Completion

1. ability
2. wait time
3. least restrictive
4. cooperative learning
5. empathetic

Matching

__c__ 1. Constructive assertiveness
__g__ 2. Higher order questioning
__b__ 3. Lower order questioning
__f__ 4. Mental modeling
__k__ 5. Responsive elaboration
__a__ 6. Self-fulfilling prophecy
__d__ 7. Student engagement

Correct terms for responses left over:
e, h, i, j [all distractors with no terms in chapter]

CHAPTER 11

Reading and Writing Instruction

LEARNING OBJECTIVES

1. Summarize the key components of skilled reading as they relate to the good information processor model from chapter one.

2. Trace the development of reading competency.

3. Discuss reading comprehension strategies taught at various age levels.

4. Compare and contrast transactional strategies instruction, whole language, and reading recovery approaches to elementary literacy instruction.

5. Explain the three major processes of skilled writing and their relationship.

6. Trace the development of writing competency.

7. Discuss several facets of writing instruction, including writer's workshop and instruction in strategies for the writing process.

8. Compare and contrast skilled reading and skilled writing as examples of expert performance.

9. Create a timeline to compare and contrast reading development and writing development.

STRENGTHENING WHAT YOU KNOW

The purpose of this chapter is to explore the nature and development of skilled reading and writing. The chapter also describes a variety of ways to foster these skills, including several instructional approaches.

Objective 1.   **Summarize the key components of skilled reading as they relate to the good information processor model from chapter one.**

1. Fill in the graphic with explanations of each component as it applies to reading

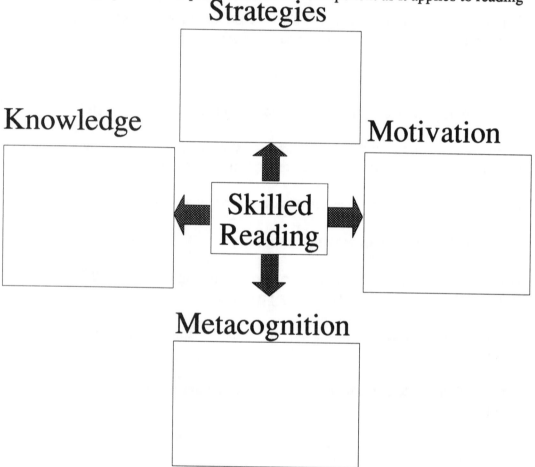

2. How good are college students when it comes to these reading components?

3. Explain how practice reading can help build all four components.

Objective 2.   **Trace the development of reading competency.**

    A.   *Emergent Literacy*

1. Define emergent literacy

2. At what ages is emergent literacy taking place?

3. Explain why storybook reading is an important part of emergent literacy by describing the interactions that take place and discussing why each would be important.

4. List other activities that can support children's literacy, and explain why each activity would be important in building literacy.

Literacy-Building Activity	Why

B. *Elementary Reading*

1. What two key aspects of reading are taught during the elementary school years?

2. Based on these two key aspects of reading, what is one of the main ways instruction changes across the elementary years?

3. Define each of the following terms and explain why each is important in decoding or learning to decode words. Be sure you make clear distinctions among the terms.

a) Phonemic awareness

b) Logographic reading

c) Phonetic cue reading

d) The cipher

e) Phonics rules

f) Orthographic patterns

g) Analogy

h) Automaticity theory

4. What kinds of comprehension instruction did Durkin find typical in her studies of elementary classrooms? What was the problem with this approach? [See Section on Reading Comprehension Strategies in textbook for discussion of how to build comprehension.]

C. *Secondary/College Reading*

1. What is the general state of literacy at the middle school, high school, and college levels?

2. What does the chapter suggest can be done to improve literacy at the secondary/college levels?

D. *Adult Reading*

1. What is the general state of adult literacy?

2. What does the chapter suggest can be done to improve adult literacy?

Objective 3. **Discuss reading comprehension strategies taught at various age levels.**

    A.   *Effective Comprehension Strategies*

Compare/contrast strategies recommended at each age. Put similar strategies in the same row. Define each strategy term and give an example.

Elementary	High School/College/Adult

B.   *Studies Skills Courses*

1. Describe a studies skills course.

2. Studies skills courses are common at the secondary and college levels. What does research show about their effectiveness?

3. What reasons are given for the results noted above?

Objective 4.   **Compare and contrast transactional strategies instruction, whole language, and reading recovery approaches to fostering elementary literacy in schools.**

A.   *Transactional Strategies Instruction*

1. What is the meaning of the term "transactional strategies instruction"?

2. Explain how transactional strategies instruction is designed to foster the components of the good information processor model (see Objective 1).

3. Based on your study of previous chapters, list any theories of development or other instructional approaches that seem consistent with the principles of transactional strategies instruction.

B.   *Whole Language*

1. Explain the term "whole language" in a way that helps you better understand this approach.

2. Explain how whole language approaches are designed to encourage the development of literacy.

3. Based on your study of previous chapters, list any theories of development or other instructional approaches that seem consistent with the principles of whole language.

4. Explain the controversy surrounding whole language approaches. Note strengths and weaknesses of whole language that make it a continuing controversy.

C.   *Reading Recovery*

1. Give an explanation of the name reading recovery that helps you understand the purpose of the approach.

2. Explain how reading recovery is designed to achieve the purpose noted above.

3. Based on your study of previous chapters, list any theories of development or other instructional approaches that seem consistent with the principles of reading recovery.

D. *Compare/Contrast Approaches*

1. Fill in the following table with features of each of the instructional approaches.

	Transactional Strategies Instruction	Whole Language	Reading Recovery
Goals			
Instructional setting			
Learning materials			
Strategies/ Skills/ Concepts Taught			
Teaching Methods			
Supporting Theories			
Related/ Similar Approaches			

Objective 6.   **Explain the three major processes of skilled writing and their relationship.**

    A.   *Recursive Nature of Writing*

1. Write the names of the 3 key writing processes in the graphic below.

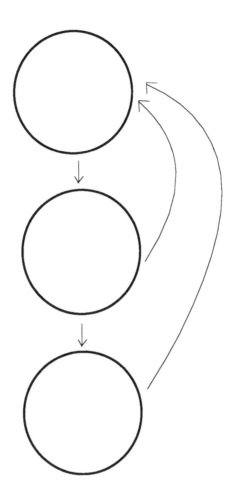

2. Arrows in the graphic above show the processes can occur at different points in writing. Give examples of how/why a writer might do each of the following.

a) Plan after composing

b) Plan after editing/revising

c) Compose after editing/revising

B. *Strong/Weak Writers and the Writing Process*

1. In the chart below, use the components of the writing process to show differences in how strong and weak writers write.

	Strong Writers	Weak Writers (See also common difficulties)
Planning		
Composing		
Editing/ Revising		

Objective 7.   **Trace the development of writing competency.**

    A.   *Emergent Writing*

1. Draw a flow chart showing 5 steps in the developmental progression that leads to writing words.

2. Which of the above types of writing would you expect from a kindergarten student?

    B.   *Elementary Writing*

1. List 5 potential problems elementary students can have with writing.

a)

b)

c)

d)

e)

C. *Secondary Writing and Beyond*

Organize the approaches college students used in planning from most effective to least effective, giving a description of each approach.

Objective 8. Discuss several facets of writing instruction, including writer's workshop and instruction in strategies for the writing process.

A. *Writer's Workshop*

List the key features of writer's workshop and explain why you think each is important.

Feature of Writer's Workshop	Why It's Important

B. *Writing Process Instruction /Strategies Instruction for Writing*

1. What is the general instructional approach for instruction in the writing process?

2. List strategies suggested for each of the phases of the writing process.

Planning strategies	
Composing strategies	
Revising strategies	

Objective 9.   **Compare and contrast skilled reading and skilled writing as examples of expert performance.**

Complete the following chart, giving specific examples of effective reading and writing that illustrate how they are instances of expert performance.

SKILLED READING	EXPERT PERFORMANCE IN GENERAL	SKILLED WRITING
	Perceive large, meaningful patterns	
	Automatized skills	
	Strong self-monitoring	
	Effectively use short-term memory	
	Effective problem-solving, representing problems at deep, principled level rather than superficial	
	Spend much of total time planning	
	Takes a long time to develop	

Objective 10. **Create a timeline to compare and contrast reading development and writing development.**

Describe the literacy behaviors and the concepts the person is learning at each developmental level.

General Age Span	Reading	Writing
Birth to preschool (emergent literacy)		
Elementary school years		
Middle school, high school, college		
Adulthood		

PRACTICE TESTS

[See answer key at the end of the chapter for correct responses.]

Multiple Choice

Circle the letter of the best response to each question.

1. Which of the following represents a component of skilled reading?
    a) Having one main way to figure things out in the text
    b) Continually stopping to check if the story makes sense
    c) Trying to just focus on the book, rather than thinking about similar personal experiences
    d) Understanding that some people have better reading ability than others

2. Which of the following represents how storybook reading is consistent with scaffolding?
    a) Adults expand on what children say about the story.
    b) Adults choose longer, more challenging stories as children become ready.
    c) Adults show that they enjoy reading and that reading is important.
    d) Adults select high-quality children's literature to read with the child.

3. Which of the following is **not** a key aspect of storybook reading to build literacy?
    a) The child points to the letters while trying to sound out words.
    b) The child asks the adult questions about the story.
    c) The adult asks the child questions about the story.
    d) The adult and child relate the story events to their own world experiences.

4. Which of the following skills often requiring instruction is one of the best predictors of success in early reading?
    a) Knowing the alphabet.
    b) Being able to use pictures to understand a story.
    c) Phonemic awareness.
    d) Logographic reading.

5. A child who reads by sounding out words based on each letter-sound relationship is using which decoding approach
    a) Phonetic cue reading
    b) Logographic reading
    c) Orthographic patterns
    d) Phonics rules

6. A student learning to read in two languages notices that there are similar root words with similar meanings across the languages. Which of the following decoding approaches has the student applied to new languages?
   a) Phonetic cue reading
   b) Logographic reading
   c) Orthographic patterns
   d) Phonics rules

7. A child knows the word MacDonald's because of the golden arches. What decoding approach is the child using?
   a) Phonetic cue reading
   b) Logographic reading
   c) Orthographic patterns
   d) Phonics rules

8. A student knows the sounds that the letters c and h make, but does not blend the "c" and "h" when trying to pronounce the word "chat." The student has not mastered which of the following?
   a) The cipher
   b) Phonics rules
   c) Phonetic cue reading
   d) Orthographic patterns.

9. Which of the following is true about phonics?
   a) Children usually pick up phonics rules without instruction.
   b) Once you learn phonics rules, they can be applied to all words.
   c) Phonics rules are so unreliable that it is not useful to learn them.
   d) Decoding by phonics rules takes up a lot of memory, making it harder to focus on understanding.

10. Which of the following comprehension strategies focuses on identifying characters, setting, plot, etc.?
    a) Summarization
    b) Story grammar training
    c) Representational imagery
    d) Mnemonic imagery

11. Which of the following comprehension strategies elaborates and transforms a story's meaning to make it more memorable?
    a) Summarization
    b) Story grammar training
    c) Representational imagery
    d) Mnemonic imagery

12. Which of the following is **not** true of reading performance as an example of expert thinking?
    a) It is useful to connect ideas from one part of the text to something else that was in the text.
    b) You need to have working memory available to remember what you read at the beginning of a sentence.
    c) It's important to recognize letters and words automatically.
    d) You should understand everything the first time you see it.

13. Of the following instructional approaches, which has the greatest focus on building decoding skills?
    a) Reading Recovery
    b) Study skills courses
    c) Transactional Strategies Instruction
    d) Whole language approaches

14. Which of the following specifically excludes explicit decoding instruction?
    a) Reading Recovery
    b) Study skills courses
    c) Transactional Strategies Instruction
    d) Whole language approaches

15. Which of the following focuses constructing understanding of a text by applying strategies as a team?
    a) Reading Recovery
    b) Study skills courses
    c) Transactional Strategies Instruction
    d) Whole language approaches

16. Which of the following would probably cover the most strategies?
    a) Reading Recovery
    b) Study skills courses
    c) Transactional Strategies Instruction
    d) Whole language approaches

17. Which of the following is an effective writing approach?
    a) Start writing down as much as you can think of about the topic.
    b) Realize that it's impossible to predict what readers already know about the topic.
    c) Closely follow a pattern you learned for forming and connecting paragraphs.
    d) Plan again after you start composing.

18. Which of the following is a common problem in revising?
    a) Not enough focus on sentence construction
    b) Unable to state ideas in meaningful sentences
    c) Not recognizing that there's a problem
    d) Not thinking about whether available information is appropriate

19. Which writing strategy would be used during the composing phase?
    a) Scan each sentence
    b) Sentence openers
    c) Story grammar
    d) Find a specific sentence in the paper that tells what the writer wants to say

20. Which of the following approaches to preparing a term paper is most effective?
    a) Note aspects of resource materials that were important or consistent with your opinion.
    b) Note ideas based on dialoguing interactively with resource materials.
    c) Write out the main ideas from your references.
    d) Think about alternative ways to write the paper and what you hope to accomplish.

Completion

Fill in each blank with the best fitting term from the chapter. Terms are used only once.

1. _____ theory that suggests that if the mind is too focused on decoding words, there is not enough brain power left to understand meaning.

2. Knowing that the letter r after a vowel sometimes changes the vowel sound is an example of a _____.

3. The set of all letter-sound relationships is called the _____.

4. Children can learn to decode by _____, guessing the sound based on a similar word.

5. Meaning develops as students and teacher work through a text together and try to understand it in _____ instruction.

6. An approach to language learning emphasizes experiences supporting natural development more than instruction is called _____.

7. An intensive tutoring program for bringing struggling readers to the level of their classmates is _____.

8. Students are involved in making choices about activities and sharing feedback with peers in the _____ approach to writing instruction.

9. Effective writing is a _____ process.

10. The process transforming rough ideas into sentences is called _____.

Matching

Match the letters of the description on the right with the corresponding numbered terms on the left. Use each description only once. Some descriptions may be left over.

_____ 1. Cipher
_____ 2. Constructive planning
_____ 3. Emergent literacy
_____ 4. Knowledge telling
_____ 5. Logographic reading
_____ 6. Orthographic pattern
_____ 7. Phonemic awareness
_____ 8. Phonetic cue reading
_____ 9. Reciprocity principle
_____ 10. Thinking aloud

a) A new way of thinking about literacy that is only beginning to be studied

b) Writers need to keep in mind what the reader knows already.

c) Using visual cues like surrounding pictures and shapes to read a word

d) Set of all letter-sound relationships

e) Research approach in which the researcher asks students to answer comprehension questions out loud rather than in writing

f) Strings of letters that hold a meaning, like prefixes and suffixes

g) Knowing words are made up of individual sounds in combination

h) Thinking about goals, criteria, problems, and strategies for writing

i) Research approach in which students tell what they are thinking and doing while they work on a task

j) Pouring out information without organizing or having a clear purpose

k) Knowing some letter-sound relationships

l) Behaviors starting at infancy that imitate and support conventional reading and writing

LEARNING STRATEGIES

Below are examples of strategies that can help you understand major chapter concepts. Use these examples to guide your own strategies.

Strategy Example #1
With a classmate, role-play a debate about the merits/problems of whole language approaches to literacy education. Even if you actually agree on the topic, you will be better able to understand both positions if each of you argues for a different side.

Strategy Example #2
Review the strategies used by more and less effective writers. Identify the strategies you use most often. Think about changes you could make in how you plan, compose, or revise that would make you a more effective writer.

Strategy Example #3
Conduct your own think-aloud study. Have a friend read a textbook out loud while you prompt them to say what they are thinking. Write down or tape what the person is saying (or tape yourself while you think aloud). Analyze the information to identify reading strategies. Do you think the strategies might be different with a different kind of reading? How could you set up a study to find out?

ANSWER KEY

Multiple Choice

Correct answers are in bold.
Comments related to other options indicate why that response is incorrect.

1. **b)**

2. **b)** Others are good practices, but this is most representative of scaffolding.

3. **a)**

4. **c)**

5. **a)**

6. **c)**

7. **b)**

8. **b)**

9. **d)** This is not to say that phonics should not be taught. The point is that the rules need to be taught to automaticity so the child can focus on other things like getting meaning.

10. **b)**

11. **d)**

12. **d)** Reading is active and strategic, often anticipating information and solving problems.

13. **a)**

14. **d)**

15. **c)**

16. **b)**

17. **d)**
     c) Strict focus on rules can interfere with communicating.

18. **c)**
     a) In contrast, too much focus on sentence construction often keeps writers from reflecting on overall meaning and the reader's needs.
     b) composing
     d) planning

19. **c)**
     a) revising
     b) planning
     d) revising

20. **d)** constructive planning
     a) TIA--true, important, I agree
     b) dialogue strategy
     c) gist and list

Completion

1. automaticity
2. phonics rule
3. cipher
4. analogy
5. transactional strategies
6. whole language
7. reading recovery
8. writer's workshop
9. recursive
10. composing

Matching

__d__ 1. Cipher
__h__ 2. Constructive planning
__l__ 3. Emergent literacy
__j__ 4. Knowledge telling
__c__ 5. Logographic reading
__f__ 6. Orthographic pattern
__g__ 7. Phonemic awareness
__k__ 8. Phonetic cue reading
__b__ 9. Reciprocity principle
__i__ 10. Thinking aloud

Correct terms for responses left over:
a) [none--distractor for emergent literacy]
e) [none--distractor for thinking aloud]

CHAPTER 12

Mathematics and Science Instruction

LEARNING OBJECTIVES

1. Compare and contrast Polya's and Schoenfeld's models of mathematics learning, showing how they relate to the good information processor model.

2. Discuss the development of mathematical understanding regarding addition and subtraction.

3. Compare and contrast the specific impacts of several effective approaches to mathematics instruction, and discuss how the methods can be integrated.

4. Discuss the roles of gender, race, and socioeconomic differences in mathematics achievement.

5. Discuss mathematics knowledge as an example of situated learning and explore the instructional implications.

6. Explore the role of misconceptions in science learning and teaching.

7. Compare and contrast the impacts of several approaches to science instruction for conceptual change.

STRENGTHENING WHAT YOU KNOW

The purpose of this chapter is to discuss advances in mathematics and science education for fostering reasoning, problem-solving skills, and conceptual change. The chapter highlights research findings on several instructional approaches for science and math.

Objective 1. **Compare and contrast Polya's and Schoenfeld's models of mathematics learning, showing how they relate to the good information processor model (presented in chapter one).**

    A.   *Polya and Schoenfeld as Models of Good Information Processing*

Complete the following chart to show how Polya and Schoenfeld each address the four components of the good information processing model. Note the additional component in Schoenfeld's model.

Polya	Good Information Processing	Schoenfeld
	Knowledge	
	Strategies	
	Metacognition	
	Motivation	

B.   Polya's Model for Problem -Solving

1. In the flow-chart below, list Polya's general model for problem-solving. In the lines to the right of each box, describe these steps based on the expanded version of Polya's model presented in the chapter.

Polya's Model   Expanded Version

2. What is a limitation of Polya's general problem-solving approach?

3. Why is an expanded version of Polya's model necessary?

C.   *Schoenfeld's Model of Mathematic Cognition*

1. What role of information processing does Schoenfeld discuss?

2. What could teachers do differently that would support good motivational beliefs about mathematics?

3. Describe Schoenfeld's instructional approach.

Objective 2. **Discuss the development of mathematical understanding regarding addition and subtraction.**

A. *Precursors to Addition and Subtraction*

Explain each of the following developmental advances in mathematical thinking in preschoolers:

1. subitizing

2. conservation of number

3. counting skills

4. count-to-cardinal transition

B. *Development of Approaches to Addition and Subtraction*

1. Explain the following strategies for solving addition problems:
a) counting on from first

b) counting on from larger

2. Which of the above addition strategies is most efficient, and why?

3. Explain the following strategies for solving subtraction problems:
a) counting down from given

b) adding on

4. How do children move from addition and subtraction strategies to using math facts?

Objective 3. **Compare and contrast the specific impacts of several effective approaches to mathematics instruction, and discuss how the methods can be integrated.**

    A.   *Comparing and Contrasting Instructional Impact*

Complete the following table, describing each instructional approach and listing the kinds of impact research has shown for each.

Instruction	Description	Impact on Learners
Cooperative Learning		
Learning from Examples		
Calculators		
Interactive Videodiscs		
Educational Television Programming		

    B.   *Complementary Methods*

1. Explain how the methods above are complementary.

Objective 4. **Discuss the roles of gender, race, and socioeconomic differences in mathematics achievement.**

    A.   *Race and Mathematics Achievement*

1. How do American students rate in mathematics compared to students in other countries?

2. What role does genetic ability have in these differences?

3. Briefly list likely reasons that Asian students perform better in mathematics than do American students.

4. Describe the mathematics achievement of minorities in the United States.

    B.   *Gender and Mathematics Achievement*

1. Compare the mathematics achievement of males and females in school, on standardized tests, and in careers.

Objective 5.   **Discuss mathematics knowledge as an example of situated learning and explore the instructional implications.**

1. In what ways is mathematics knowledge "situated"?

2. What are the implications of situated mathematics learning for educators?

Objective 6.   **Explore the role of misconceptions in science learning and teaching.**

    A.   *Piaget's Views of the Role of Misconceptions*

1. Describe Piaget's views about the role of children's science misconceptions in their science learning.

2. What aspect of Piaget's views has been found incorrect? Explain with supporting evidence.

3. What aspects of Piaget's views has been used to develop instruction for changing science misconceptions?

Objective 7. **Compare and contrast the impacts of several approaches to science instruction for conceptual change.**

    A.   *Comparing and Contrasting Instructional Impact*

1. Explain the conceptual change strategy

2. Complete the following table, describing each instructional approach and listing the kinds of impact research has shown for each.

Instruction	Description/Rationale	Impact on Learners
Conceptual Conflict and Accommodation Model		
Analogies		
Refutation Text		
Multiple and Alternative Representations		
Discovery Learning		
Socially-supported, collaborative construction of science concepts		

3. List several types of alternative representations that can help students in science.

B. *Discovery Learning and Socially-Supported, Collaborative Construction of Science Concepts*

1. Why do many science educators want students to engage in discovery learning, despite its drawbacks?

2. Explain how socially-supported, collaborative construction of science concepts is different from pure discovery learning.

3. Explain the roles thinking aloud and scaffolding can play in socially-interactive science instruction.

4. What are the potential problems of students collaborating in science instruction?

PRACTICE TESTS

[See answer key at the end of the chapter for correct responses.]

Multiple Choice

Circle the letter of the best response to each question.

1. Which of the following is a limitation of Polya's problem-solving model?
    a)   It is a general approach designed to work across situations.
    b)   It is too much like the good information processing model to be useful.
    c)   The steps don't help people solve problems better.
    d)   It does not address the need for metacognition.

2. Which of the following is an example of Polya's step "Devise a plan for solving the problem"?
    a)   Compare the answer obtained with an estimated answer.
    b)   Identify the important information, such as in a list or table.
    c)   Identify a familiar pattern that you have seen in a previous problem.
    d)   Create a representation of the problem with drawings or objects.

3. Which component of Schoenfeld's model is **not** addressed in Polya's model?
    a)   Monitoring
    b)   Instruction
    c)   Strategies
    d)   Motivation and attitudes

4. According to Schoenfeld, which of the following is the most helpful belief for learners to have about mathematics?
    a)   To be a good math student, I should be able to answer the teacher's questions right away.
    b)   If I really want to do well in this math class, I should spend my homework time remembering the formulas that I have to use on the test.
    c)   I guess I'm just one of those people who is naturally good at math.
    d)   I could create my own way to solve this problem, even though it's not the way the teacher taught me.

5. Which instructional approach is most consistent with Schoenfeld's approach?
   a) Discovery learning
   b) Cooperative learning
   c) Reciprocal teaching
   d) Direct explanation

6. Which of the following is true of counting in preschool children?
   a) It is limited to being able to say numbers in sequence.
   b) By kindergarten, most children can only count objects up to about 10.
   c) Children arrange objects for counting but often skip one or count one twice.
   d) Counting objects usually becomes automatic by age 5.

7. A child is given the problem 3 + 6=?. The child says, "Three. Four, five, six, seven, eight, nine. The answer is nine." Which of the following would be a **more efficient** strategy for solving the same problem?
   a) Counting on the fingers
   b) Counting down from given
   c) Counting on from larger
   d) Adding on

8. Which of the following approaches would a fourth-grade student most likely use to solve a subtraction problem?
   a) Memorize the answer
   b) Adding on
   c) Counting on from larger
   d) Counting down from given

9. Which of the following mathematics instruction approaches creates the best opportunities to gather information needed to understand complex relationships in a rich, realistic context?
   a) Learning from examples
   b) Interactive videodisc
   c) Cooperative learning
   d) Calculators

10. Which of the following mathematics instruction approaches helps reduce math anxiety?
    a) Learning from examples
    b) Interactive videodisc
    c) Cooperative learning
    d) Calculators

11. Which of the following mathematics instruction helps students quickly learn procedures for solving a particular type of problem?
    a) Learning from examples
    b) Interactive videodisc
    c) Cooperative learning
    d) Calculators

12. Which of the following is true of across the instructional approaches for mathematics discussed in the chapter?
    a) A teacher should chose one of the approached and stick with it.
    b) They don't focus on real-life problems.
    c) They require students to create innovative solutions.
    d) They mainly give students information about how to solve problems.

13. Which of the following is true of cooperative learning?
    a) Girls learn best in all-girl cooperative groups.
    b) Active participation is best when the students in the group cover the full ability range in the class.
    c) When one student presents an incorrect solution, the other students most often catch the mistake.
    d) Sharing ideas about solutions helps students value the approach to solving the problem instead of just getting the final right answer.

14. Which of the following differences is **not** a likely reason that Asian students perform better in mathematics than do American students?
    a) Genetic differences in spatial and mathematical reasoning
    b) Time students spend working on mathematics
    c) Amount of focus on the processes for solving problems
    d) Teachers' time in the classroom

15. Which of the following is true?
    a) American racial minorities perform as well in mathematics as members of the majority culture.
    b) Poverty has little impact on mathematics achievement.
    c) Girls get better math grades than boys.
    d) The minority/majority mathematics acheivement gap is getting wider.

16. In which of the following ways was Piaget's views on science learning incorrect?
    a) People often have misconceptions about scientific matters.
    b) Children often believe inanimate objects are alive.
    c) Nonscientific ideas change in concrete or formal operations.
    d) Cognitive conflict can be used to help students change their misconceptions about science.

17. Which science instruction approach simply tells students common misconceptions?
    a) analogies
    b) conceptual conflict and accommodation model
    c) multiple representations
    d) refutation text

18. Which of the following science instruction approaches parallels one way scientists often think about concepts?
    a) discovery learning
    b) conceptual conflict and accommodation model
    c) multiple representations
    d) refutation text

19. Which science instruction approach has students express their own views about scientific principles to make them aware of possible misconceptions?
    a) analogies
    b) conceptual conflict and accommodation model
    c) multiple representations
    d) refutation text

20. Choose the instructional approach that has this drawback: Students often cannot tell when they are closer to understanding the concept or problem.
    a) discovery learning
    b) conceptual conflict and accommodation model
    c) multiple representations
    d) refutation text

21. Which of the following is **not** among the potential problems of using analogies?
    a) Students easily go back to their old ways of thinking.
    b) Students misunderstand relationships between the analogy and concept.
    c) Students are unfamiliar with how the analogy itself works.
    d) The analogy breaks down and leads to misconceptions.

Completion

Fill in each blank with the best fitting term from the chapter. Terms are used only once.

1.   Polya's strategies had a greater impact on students as their _____ increased.

2.   Schoenfeld points out that well organized knowledge of mathematics reduces _____ demands when solving problems.

3.   One reason metacognition is so important in Schoenfeld's model is that students often have consistent _____ when solving a certain kind of problem.

4.   As Piaget suggested about learning, students reading science texts that contrast with their existing beliefs often try to _____ information.

5.   A foundation for many approaches to science instruction is the _____ strategy.

6.   An approach to instruction that capitalizes on benifits of discovery learning while trying to offset its problems is referred to as _____.

Matching

Match the letters of the description on the right with the corresponding numbered terms on the left. Use each description only once. Some descriptions may be left over.

_____ 1. Adding on
_____ 2. Animism
_____ 3. Conservation of number
_____ 4. Counting down from given
_____ 5. Counting on from first
_____ 6. Counting on from larger
_____ 7. Count-to-cardinal transition
_____ 8. Subitizing

a) Ability to identify the number of objects in a small set

b) Subtraction strategy in which the child counts backwards to reach the answer

c) Arranging objects from small to large and starting to count from the biggest one

d) Believing objects are alive even though they are not living things.

e) Children notice when an object is removed, even though the objects span the same length.

f) Starting with the first object in a set and counting all the objects

g) Also called the "MIN" strategy because it minimizes the number of counts required to solve the problem

h) Subtraction strategy of seeing how many counts it takes to get from the smaller number to the larger one

i) Problem is 2 + 8. Child says, "Two. Three, four, five, six, seven, eight, nine, ten."

j) Knowing that the last number when counting objects is the number of objects in the set

LEARNING STRATEGIES

Below are examples of strategies that can help you understand major chapter concepts. Use these examples to guide your own strategies.

Strategy Example #1
To illustrate how the mathematics instruction approaches in the chapter are complementary, rather than competing methods, choose at least three of the approaches and describe a lesson that would integrate them.

Strategy Example #2
Brainstorm possible reasons for mathematics achievement gaps for females, for cultural minorities, and for students from low socioeconomic backgrounds. Choose two or three causes that you consider the most likely, and think of possible ways to deal with those causes.

Strategy Example #3
Which of the science instruction approaches are complementary, and which are competing approaches? Sketch ideas for a lesson that would integrate complementary approaches. For competing approaches, have a mock debate with a classmate, in which each of you argues why your approach is better. Jot down notes of what the debate reveals about the approaches.

ANSWER KEY

Multiple Choice

Correct answers are in bold.
Comments related to other options indicate why that response is incorrect.

1. **a)** Domain-specific approaches are also needed for efficient solutions to common problems.

2. **c)**

3. **b)**

4. **d)**
   c) Although this is a positive attitude, it represents attributing success to ability rather than effort.

5. **d)**

6. **c)**

7. **c)**
   a) not necessarily a faster approach
   b) a subtraction strategy

8. **a)**

9. **b)**

10. **c)**

11. **a)**

12. **c)**
    a) They are complementary and are effective in combination.
    d) Students have to be reflective and come up with their own approaches.

13. **d)**
    b) some diversity, but not the full range, is best
    c) Often, good solutions are rejected and poor ones accepted, a potential drawback of this approach.

14. **a)**

15. **c)** Boys do better only on standardized tests.

16. **c)**

17. **d)**

18. **b)** Analogies, a type of representation, also parallel how scientists think.

19. **b)**

20. **a)**

21. **a)**

Completion

1. grade level
2. short-term capacity/working memory
3. errors / bugs
4. accomodate
5. conceptual change
6. socially supported, collaborative construction of science concepts

Matching

__h__ 1. Adding on
__d__ 2. Animism
__e__ 3. Conservation of number
__b__ 4. Counting down from given
__i__ 5. Counting on from first
__g__ 6. Counting on from larger
__j__ 7. Count-to-cardinal transition
__a__ 8. Subitizing

Correct terms for responses left over:
c) [none--distractor]
f) [none--distractor]

CHAPTER 13

Traditional Perspectives on Intelligence and Academic Competence

LEARNING OBJECTIVES

1. Compare and contrast theories of intelligence.

2. Discuss changes in intellectual competencies across the lifespan.

3. Discuss individual differences in intelligence, especially the roles of heredity and environment in determining intelligence.

4. Explain key criteria for evaluating tests--reliability and validity, and discuss factors that influence them.

5. Be able to interpret standardized test scores in several formats.

6. Compare and contrast several common intelligence tests.

7. Discuss the controversies of bias in mental testing, as well as potential sources of test bias.

8. Discuss instructional interventions designed to increase intelligence.

STRENGTHENING WHAT YOU KNOW

The purpose of this chapter is to discuss issues of assessment on intelligence and academic competence. The chapter discusses theories of intelligence, common intelligence tests, criteria for evaluating tests, interpreting test scores, and possible problems with intelligence tests.

Objective 1.   **Compare and contrast theories of intelligence**.

    A.   *Comparing Theories of Intelligence*

THEORIST/ THEORY OF INTELLIGENCE	How many factors?	Describe the factors and their relationship (if information is available)
Spearman		
Thurstone		
Guilford		
Cattell and Horn's Factors of Intelligence		
Sternberg's Triarchic Theory of Intelligence		
Gardner's Multiple Intelligences		

B.   *Cattell and Horn's Factors of Intelligence*

1. Define the following terms and explain their role in Cattell and Horn's theory:

fluid intelligence

crystallized intelligence

quantitative intelligence

short-term memory

long-term memory

visual processing

auditory processing

processing speed

correct decision speed

2. In what ways is this theory similar to the view of the Good Information Processor?

C.   *Sternberg's Triarchic Theory of Intelligence*
1. In the graphic below, describe Sternberg's three sub-theories of intelligence.

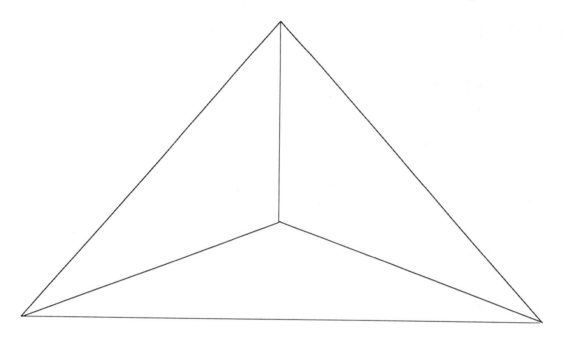

2. Explain how the following components correspond to the Good Information Processor model:

meta-components

performance components

knowledge acquisition components

3. What similarities and differences do you see between Sternberg's theory and Cattell and Horn's theory?

D.   *Gardner's Theory of Multiple Intelligences*

1. What are the instructional implications of Gardner's theory?

Objective 2. **Discuss changes in intellectual competencies across the lifespan.**

A. *Stability of IQ*

1. How stable is IQ from childhood to early adulthood?

2. Some studies suggest a big drop in IQ as adults age, whereas other studies show a less dramatic change? What main factor accounts for these different findings? Which finding is more trustworthy, and why?

B. *Impact of Aging on Different Aspects of Intelligence*

1. Fill in the chart to compare how knowledge and processing abilities change with aging. Explain the reasons behind these differences.

Crystallized Intelligence--Knowledge	Fluid Intelligence--Processing Ability

2. What aspects of the good information processing model are affected by age, and how?

3. What aspects of the good information processing model are **not** affected by age?

Objective 3.   **Discuss individual differences in intelligence, especially the roles of heredity and environment in determining intelligence.**

1. If children's intelligence tends to be similar to their parents' intelligence level, why is it hard to say if intelligence is inherited?

A.   *Studies on the Role of Heredity and Environment*

1. Compare the IQ correlations from several types of studies. Explain what each kind of study can add to understanding the role of heredity and environment.

	IQ Correlation	Importance of Comparison
Identical twins reared together		
Identical twins reared apart		
Fraternal twins reared together		
Parents and their biological children reared in their homes		
Biological parents and their children who do not live with them		
Parents and their adopted children		
Biological siblings reared together		
Biological siblings reared apart		
Biological half-siblings		

2. Summarize the results of these studies in statements about the role of heredity and environment on intelligence.

Objective 4. **Explain key criteria for evaluating tests--reliability and validity, and discuss factors that influence them.**

A. *Reliability*

1. Define reliability.

2. Explain why reliability is the first criteria of a test. In what way do other criteria depend on reliability?

3. Explain the meaning behind the formula "observed score = true score + error."

4. List possible sources of testing error.

5. How does standardization help reduce error and increase reliability?

6. In the table below, explain the various approaches used to estimate a test's reliability, including how each is measured. Note points of comparison and contrast.

test-retest reliability	
alternate-forms reliability	
split-half reliability	
coefficient alpha	

      B.   *Validity*

1. Define validity.

2. Explain how a test can be reliable without being valid. (Can the reverse be true?)

3. In the table below, explain three common types of validity, including how each is measured. Note points of comparison and contrast.

Construct Validity	
Content Validity	
Criterion Validity	

Objective 5.   Be able to interpret standardized test scores in several formats.
    A.   *Standardized Tests*
1. Define "standardized test" and explain the purpose of testing this way.

2. Define test norms and explain how they are used.

    B.   *Standardized Test Scores*
1. a) Define the "mean" of a set of scores.

b) Explain why it is useful to compare a raw score to a measure of central tendency.

2. a) What useful information does the standard deviation provide?

b) What is the relationship between the mean and the standard deviation?

3.   Compare and contrast the following three types of scores.

	Z-score	Stanine	Percentile rank
How is it calculated?			
What does it mean?			
Strengths			
Weaknesses			

Objective 6. **Compare and contrast several common intelligence tests.**

A. *Purposes of IQ Tests*

Describe the purposes and usefulness of intelligence tests.

B. *Comparisons of Widely Used Individual Intelligence Assessments*

	Ages	View of Intelligence	Description of Scales / Tasks	Reliability	Validity	strengths/ weaknesses
WAIS-III						
WISC-III						
WPPSI-R						
Stanford-Binet						
K-ABC						

C.   *Intelligence Tests for Special Populations*

1. Compare the following IQ tests for infants by filling the table with whatever information is available in the chapter.

	Ages	Types of Tasks	What does it predict?	Does it predict childhood intelligence?
Brazleton Neonatal Behavioral Assessment Scale				
Bayley Scales of Infant Development				
Visual Habituation Paradigm				

2. List and describe four intelligence tests for individuals with special characteristics:

a)

b)

c)

d)

D.   *Group-Administered Intelligence Tests*

1. List the strengths and weaknesses of group-administered intelligence tests:
STRENGTHS   WEAKNESSES

2. What is one way that group-administered measures should **not** be used? Why?

Objective 7. **Discuss the controversies of bias in mental testing, as well as potential sources of test bias.**

    A.   *IQ and Cultural Differences*

1. How do IQ scores compare across different cultural groups?

2. What social factor accounts, in part, for these differences?

    B.   *Sources of Bias in Mental Testing*

Summarize five possible sources of test bias. How much of a threat does each pose, and how can it be overcome?:

1)

2)

3)

4)

5)

    C.   *Court Rulings on Test Use*

1. Summarize the information on how U. S. Courts are responding to cases of test bias.

Objective 8. **Discuss instructional interventions designed to increase intelligence.**

A. *Genotypes and Phenotypes*

1. Define each of the following terms, and explain the role each can play in intelligence. Give examples to illustrate.

Genotype:

Phenotype:

Reaction range:

B. *Interventions for Increasing Intelligence*

1. Describe the kinds of activities and interventions that have been used in efforts to improve children's IQ scores.

2. According to the chapter, can such approaches really improve a person's IQ?

PRACTICE TESTS

[See answer key at the end of the chapter for correct responses.]

Multiple Choice

Circle the letter of the best response to each question.

1.   A first requirement of a test is
     a)   Validity
     b)   Reliability
     c)   Standardization
     d)   Norms

2.   Which of the following would be measured through correlations with related existing tests?
     a)   Reliability
     b)   Content validity
     c)   Construct validity
     d)   Criterion validity

3.   An increase in which of the following will increase reliability?
     a)   Time between testings
     b)   Test-taker anxiety
     c)   Test length
     d)   Error

4.   Which of the following factors would effect a test's validity?
     a)   Whether scoring is more objective or more open to interpretation
     b)   What group the person's score is being compared with
     c)   Whether the question is vague or confusing
     d)   What the test results are being used for

5.   Which of the following is an average that makes up for problems in simply comparing odd items with even items?
     a)   test-retest reliability
     b)   alternate-forms reliability
     c)   coefficient alpha
     d)   split-half reliability

6. Which of the following is sometimes known as predictive validity or concurrent validity?
   a) Reliability
   b) Content validity
   c) Construct validity
   d) Criterion validity

7. Which of the following theorists highlights how individuals adapt to their social and cultural environment?
   a) Cattell and Horn
   b) Gardner
   c) Spearman
   d) Sternberg

8. Which of the following theorists suggest(s) that each person has a specific set of strengths in various areas of intelligence, determined genetically?
   a) Cattell and Horn
   b) Gardner
   c) Spearman
   d) Sternberg

9. Which of the following theorists suggest(s) that intelligence involves coordinating several mental processes?
   a) Cattell and Horn
   b) Gardner
   c) Spearman
   d) Sternberg

10. Which of the following is true about the history of IQ tests?
    a) Early tests that measured general intelligence (g) were poor at making discriminations in ability.
    b) IQ tests were originally used to identify children with mental retardation.
    c) IQ is negatively correlated with occupational success.
    d) IQ scores cannot predict important life outcomes.

11. Which of the following is **not** part of Cattell and Horn's theory of intelligence?
    a) language skills
    b) thinking speed
    c) memory
    d) quantitative skills

12. Which of the following intelligence tests is designed to be used across the widest span of ages?
   a) K-ABC (Kaufman)
   b) WAIS-III
   c) WISC-III
   d) WPPSI-R
   e) Stanford-Binet

13. Which of the following measures mental processes, not just outcomes?
   a) K-ABC (Kaufman)
   b) WAIS-III
   c) WISC-III
   d) WPPSI-R
   e) Stanford-Binet

14. Which of the following was specifically developed for children ages 4-6.5?
   a) K-ABC (Kaufman)
   b) WAIS-III
   c) WISC-III
   d) WPPSI-R
   e) Stanford-Binet

15. Of the following, which test of infant intelligence best predicts childhood intelligence?
   a) Bayley Scales of Infant Development
   b) Brazleton Neonatal Behavioral Assessment Scale
   c) Raven's Progressive Matrices
   d) Visual Habituation Paradigm

16. Which of the following types of measures is **least appropriate** for making instructional decisions?
   a) Group administered measures
   b) Non-verbal measures
   c) Verbal measures
   d) Visual habituation measures

17. If you want to know how many students did worse than you on a test, which of the following types of scores would you request?
   a) Stanine
   b) Z-score
   c) Standard deviation
   d) Percentile

18. Which of the following decreases as adults age?
    a)   Crystallized intelligence
    b)   Processing capacity
    c)   Monitoring abilities
    d)   Knowledge

19. Which of the following statements is best supported by the findings of twin studies and adoption studies?
    a)   Because identical twins have similar IQs, intelligence is hereditary.
    b)   The IQ of an adoptive family has little impact on a child's IQ.
    c)   About half the variability in IQ is due to heredity.
    d)   You can't tell anything from these studies because there are too many alternative explanations.

20. Which of the following is least likely to be the source of test bias?
    a)   The test-taker's cultural group is under-represented in the norm sample.
    b)   The person giving the test has a different racial background than person taking the test.
    c)   The test could be good at predicting the success of one group, but poor at predicting the success of another group.
    d)   The test may focus on concepts that are not familiar to all cultural groups.

21. Sally's score is in the 41st percentile on a test. What does this mean?
    a)   Sally answered 41 out of every 100 items correctly.
    b)   Sally has low ability.
    c)   41% of the people did the same as or better than Sally did on the test.
    d)   41% of the people did the same as or worse than Sally did on the test.

22. Which of the following represents the most critical problem with interventions that are designed to increase IQ?
    a)   They don't make a difference in students' IQ scores.
    b)   They don't start early enough in the child's life.
    c)   They don't follow through long enough.
    d)   They are not powerful enough to overcome genetic factors.

Completion

Fill in each blank with the best fitting term from the chapter. Terms are used only once.

1. An intelligence measure based on how long an infant looks at familiar and unfamiliar faces is called _____.

2. A score on a test that reflects the person's actual ability as well as chance factors that affect performance is called the _____.

3. A correlation of a person's scores between two versions of the same test is called _____ reliability.

4. Newborns' reflexes are evaluated as a sign of intelligence in the _____.

5. A _____ test is given in a controlled situation so every test taker has the same test experience.

6. Genetic heritage is expressed in one of many possible _____, and the "reaction range" of these possibilities depends heavily on one's environment.

7. A comparison of the same person's score on two test occasions is _____ reliability.

8. The internal consistency of a test is measured through _____ reliability procedures.

9. An individual's test score is often compared to a _____, the typical performance level for a well-defined group of people.

10. When scores are clustered around the mean, with fewer extreme high and low scores, they are following a _____ distribution.

Matching

Match the letters of the description on the right with the corresponding numbered terms on the left. Use each description only once. Some descriptions may be left over.

_____ 1. Construct validity
_____ 2. Content validity
_____ 3. Criterion validity
_____ 4. Mean
_____ 5. Percentile
_____ 6. Reliability
_____ 7. Standard deviation
_____ 8. Stanine
_____ 9. Validity
_____ 10. Z-score

a) Tells how many standard deviations a score is away from the mean
b) Test measures what it is intended to measure
c) Represents a group of raw scores that fall in a certain region of the normal distribution
d) Measure of the strength of relationship between two psychological constructs
e) Measure of whether the test appropriately measures the specific psychological variable it was designed to measure
f) A measure of central tendency
g) Test measures consistently
h) Tells a person's position within a group
i) Analysis of correlations between items, to see if they are testing the same thing
j) Shows whether the test makes the kind of distinctions among people that it is supposed to make
k) Measure of variation in scores
l) Focuses on whether a test covers the topics it is supposed to, in the appropriate proportions

LEARNING STRATEGIES

Below are examples of strategies that can help you understand major chapter concepts. Use these examples to guide your own strategies.

Strategy Example #1
Go to the library and find a copy of the <u>Mental Measurements Yearbook</u> or <u>Tests in Print</u>. Study the information for one test and answer the "Questions to Ask Yourself" to evaluate the test.

Strategy Example #2
Calculate your personal mean and standard deviation for your grades in a course. If no number grades are given, convert letter grades to numbers (e.g., A-4 B-3 C-2 D-1 F-0). What do the mean and standard deviation tell you about your grades in the class? Choose a particular assignment, and convert the score to a z-score, stanine, and percentile. How are the meanings of these scores similar to/different from standardized test scores comparing across people?

Strategy Example #3
Draw a diagram to show the relationships among the following:
genotype, phenotype, reaction range, environment, and intelligence.

ANSWER KEY

Multiple Choice

Correct answers are in bold.
Comments related to other options indicate why that response is incorrect.

1. **b)**
   c) used to improve reliability

2. **c)**

3. **c)**

4. **d)**
   a) reliability issue
   b) norm issue
   c) reliability issue

5. **c)**

6. **d)**

7. **d)**

8. **b)**

9. **a)**

10. **b)**

11. **a)** Gardner's multiple intelligences

12. **e)**

13. **a)**

14. **d)**

15. **d)**
    c) nonverbal test designed for children, not infants

16. **a)**

17. **d)**

18. **b)**
    a) increases
    d) knowledge is related to crystallized intelligence

19. **c)**
    d) Although there are alternative explanations, a wide variety of studies lead to the same general conclusion.

20. **b)** knowing the examiner can cause bias; examiner's race does not

21. **d)**

22. **c)**

Completion

1. visual habituation
2. observed score
3. alternate-forms
4. Brazleton Neonatal Behavioral Assessment Scale
5. standardized
6. phenotypes
7. test-retest
8. split-half
9. norm
10. normal

Matching

__e__ 1. Construct validity
__l__ 2. Content validity
__j__ 3. Criterion validity
__f__ 4. Mean
__h__ 5. Percentile
__g__ 6. Reliability
__k__ 7. Standard deviation
__c__ 8. Stanine
__b__ 9. Validity
__a__ 10. Z-score

Correct terms for responses left over:
d) correlation
i) factor analysis

CHAPTER 14

Alternative Assessments of Academic Competence

LEARNING OBJECTIVES

1. Describe the National Assessment of Educational Progress, and discuss its uses.

2. Describe performance assessments, and discuss practical considerations for their use.

3. Describe portfolio assessments, their role in national and state-wide testing, and practical considerations for their use.

4. Describe dynamic assessment, and discuss evaluations and instructional implications of this testing approach.

5. Compare and contrast four models of curriculum-based assessment.

6. Compare and contrast forms of biological-based assessment.

7. Compare and contrast the types of assessment in the chapter.

STRENGTHENING WHAT YOU KNOW

The purpose of this chapter is to describe testing programs that affect public policy and to explore alternative approaches to evaluating students' academic competence.

Objective 1.   **Describe the National Assessment of Educational Progress, and discuss its uses.**

     A.   *Purposes of the NAEP*

1. Explain the reasons for having nation-wide and statewide tests.

2. Describe ways the results of the NAEP are used (and often misused).

USES	MISUSES

3. What subject areas does the test cover?

4. List common criticisms of the NAEP.

5. List teacher's typical concerns about state and national testing in general.

Objective 2. **Describe performance assessments, and discuss practical considerations for their use.**

A. *Focus of Performance Assessments*

1. How is the focus of performance assessments different from the focus of traditional assessments?

2. From an educational standpoint, list several reasons performance assessment might be better than tests that focus only on a correct answer.

B. *Considerations in Developing and Implementing Performance Assessment*

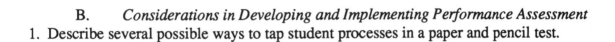

1. Describe several possible ways to tap student processes in a paper and pencil test.

2. Describe various methods for scoring performance assessments.

3. List several practical considerations for implementing performance assessments.

Objective 3. **Describe portfolio assessments, their role in national and state-wide testing, and practical considerations for their use.**

    A.   *Portfolio Assessment*

1. In what ways is portfolio assessment different from other forms of performance assessment?

3. Briefly describe some of the different possible purposes of portfolios.

    B.   *Portfolios in National and State-wide Testing*

1. Explain the current status of portfolio assessments in the NAEP.

2. Explain the current status of portfolio assessments in statewide testing.

3. What evidence suggests that portfolios may be useful in national and statewide testing?

    C.   *Reliability and Validity*

1. In the chart below, list contrasting arguments and information from the chapter regarding the reliability and validity of portfolio assessments.

	Reliability	Validity
+		
-		

Objective 4. **Describe dynamic assessment, and discuss evaluations and instructional implications of this testing approach.**

    A.   *Dynamic Assessment*

1. Describe a dynamic assessment.

2. Why might dynamic assessment be considered a more fair test of intelligence?

    B.   *Evaluations of Dynamic Assessment*

1. What information is there regarding the reliability of dynamic assessment?

2. What information is there about the validity of dynamic assessment?

    C.   *Instructional Implications of Dynamic Assessment*

1. What populations would probably benefit the most from dynamic assessment? Why?

2. Explain how dynamic assessment could be used to facilitate instruction.

Objective 5. **Compare and contrast four models of curriculum-based assessment.**

A. *Purposes of Curriculum-Based Assessment*

1. Explain the general purpose and use of curriculum-based assessment. In what way is curriculum-based assessment better suited to this purpose than other testing approaches?

B. *Models of Curriculum-based Assessment*

1. Complete the following table, highlighting features that distinguish among the models.

	Description of procedures	Specific Purposes / Uses
Curriculum-Based Assessment for Curriculum Design		
Curriculum-Based Measurement		
Criterion-Referenced Models		
Curriculum-Based Evaluation		

2. List the similarities across the models of curriculum-based assessment.

3. What is known about reliability and validity of these approaches?

Objective 6.   Compare and contrast forms of biological-based assessment.

A.   *Brain Structures and Functions*

1. Define brain "structures."

2. Define brain "functions."

3. Describe the relationship between brain structures and brain functions.

B.   *Types of Biological Assessment*

	Description	Current/ Potential Uses & Educational Implications	Relationship to IQ
Neurological Test Batteries			
Brain imaging PET Scan			
MRI			
Psycho-physiological Measures			

Objective 7.   **Compare and contrast the types of assessment in the chapter.**

	Description	Uses / Implications	Strengths	Weaknesses
NAEP				
Performance Assessment				
Portfolio Assessment				
Dynamic Assessment				
Curriculum-Based Assessment				
Biologically-Based Assessment				

PRACTICE TESTS

[See answer key at the end of the chapter for correct responses.]

Multiple Choice

Circle the letter of the best response to each question.

1. Which of the following is true of the NAEP?
    a) The same subject areas are tested every year.
    b) It tests all children in the United States.
    c) The test asks students questions about their home environment.
    d) It is designed so that there will not be achievement differences among cultural groups.

2. Which of the following represents a common misuse of the NAEP?
    a) Determining federal and state education policies
    b) Looking for characteristics of instruction that promote achievement
    c) Tracking trends in student performance over the years
    d) Comparing schools to show how well they are doing

3. Which of the following is **not** a common criticism of the NAEP?
    a) It needs to test deeper understanding.
    b) It is not good for comparing across states or across schools.
    c) It unfairly makes the younger generation look bad.
    d) It lacks content validity.

4. Which of the following is true of paper-and-pencil performance assessments?
    a) They are scored mostly for the final answer, with extra credit given for the solution process
    b) They include problems just as complex as real-world problems
    c) They evaluate students' metacognitive awareness
    d) They are simple to develop and administer

5. Which of the following types of assessment teaches new information and evaluates how well students learn it?
    a) Biologically-based assessment
    b) Curriculum-based assessment
    c) Dynamic assessment
    d) Portfolio assessment

6. Mrs. Clemens, a first-grade teacher, is interested in seeing how her students' understanding of concepts changes as they progress through the grades. Which type of assessment should Mrs. Clemens recommend to the school board for meeting this purpose?
   a) Biologically-based assessment
   b) Curriculum-based assessment
   c) Dynamic assessment
   d) Portfolio assessment

7. Which type of assessment is being considered for national and statewide testing?
   a) Biologically-based assessment
   b) Curriculum-based assessment
   c) Dynamic assessment
   d) Portfolio assessment

8. Dr. Koi is a superintendent who is guiding the development of instructional policy for her school system. Which of the following forms of assessment is most likely to help her make informed decisions?
   a) Portfolio assessment
   b) The NAEP
   c) Dynamic assessment
   d) Curriculum-based assessment

9. Dr. Gallego is the principal of an inner-city school in a high poverty area. He needs a test that will show differences among students and help teachers address students' specific learning needs. Which type of test would be most helpful?
   a) The NAEP
   b) Dynamic assessment
   c) Curriculum-based assessment
   d) Biologically-based assessment

10. Dr. Semely is a school psychologist who serves many students suffering from disabling diseases. To guide teachers in providing appropriate instruction, she needs information about the students' specific mental abilities and limitations. Which type of test would be most helpful?
    a) PET scan
    b) Dynamic assessment
    c) Psychophysiological measures
    d) Neuropsychological test batteries

11. Which of the following is true of dynamic assessment?
    a) It is severely unreliable.
    b) It measures learning speed.
    c) It is a poor predictor of academic achievement.
    d) It measures the skills and knowledge a person has acquired.

12. Which type of measure is based on neural efficiency theory?
    a) Choice reaction time
    b) PET scan
    c) Being able to clap a rhythm
    d) Dynamic assessment

13. Which type of assessment is most closely related to mastery learning?
    a) Portfolio assessment
    b) The NAEP
    c) Dynamic assessment
    d) Curriculum-based assessment

14. You receive a letter from your child's school, asking your permission to give your child a test that includes the following tasks: remembering nonsense words, clapping a rhythm, and reading and explaining a paragraph. Based on this information, what kind of assessment is this, most likely?
    a) Curriculum-based
    b) Dynamic
    c) Neuropsychological
    d) Performance/portfolio

15. Your child comes home with test results. The test shows a set of goals, with a graph of how your child has progressed on each goal. Based on this information, what kind of assessment is this, most likely?
    a) Curriculum-based
    b) Dynamic
    c) Neuropsychological
    d) Performance/portfolio

16. Which of the following is true of brain imaging techniques?
    a) They give little information about mental functioning.
    b) They show that smart people have more brain activity when they are doing a task.
    c) They show that each region of the brain performs a specific job and cannot do the work of another part of the brain.
    d) They can show what parts of the brain are working at a given time.

17. Which of the following is true of Neuropsychological tests?
    a) They are more for medical diagnosis than education.
    b) The tasks do not resemble typical academic tasks.
    c) There are many tests for specific brain functions.
    d) Only a few children require this kind of assessment.

18. Which of the following measures is already being used to identify individual differences in basic functioning that may relate to general intelligence?
    a) Brain imaging
    b) Dynamic assessment
    c) Psychophysiological measures
    d) Neuropsychological test batteries

19. Which of the following is a correct evaluation of performance assessment?
    a) The detailed nature of performance assessment gives a reliable estimate of performance, even though there are only a few items.
    b) Scoring was reliable across raters.
    c) Student performance was reliable across tasks.
    d) The validity of performance assessment is it's strength, even though reliability might be low.

20. Which of the following represents a view teachers often have about national testing?
    a) The tests give students outside motivation to learn content.
    b) Teachers need administrative support to set aside more time to better prepare students for these tests.
    c) Negative feedback of tests can lead to learned helplessness in teachers.
    d) National tests have little impact on local curriculum.

Matching

Match the letters of the description on the right with the corresponding numbered terms on the left. Use each description only once. Some descriptions may be left over.

_____ 1. Authentic assessment
_____ 2. Choice reaction time
_____ 3. Function
_____ 4. MRI
_____ 5. Neural efficiency
_____ 6. PET scan
_____ 7. Portfolio
_____ 8. Structure

a) Information about the processes disrupted because of a brain disorder or injury
b) Set of items intended to reflect students' learning
c) After injecting or inhaling a radioactive substance, the person's brain glows so that activity in different structure of the brain can be measured.
d) A test of learning potential.
e) Information about parts of the brain involved in a brain disorder or injury
f) A nationwide test of academic achievement
g) Fast mental processes take up less capacity than slower processes.
h) Uses magnetic fields to explore the structure of the brain
i) Measure in which the test-taker pushes a button when a light goes on
j) Focusing on real-world problems

LEARNING STRATEGIES

Below are examples of strategies that can help you understand major chapter concepts. Use these examples to guide your own strategies.

Strategy Example #1

Brainstorm possible ways to score performance assessments. How easy or difficult do you think it would be to get strong reliability using each scoring approach? Why?

Strategy Example #2

Working with a classmate or group of classmates, have each person choose a different form of assessment. Prepare for and conduct a mock debate about which kind of assessment should be used in your state. Be aware of the weaknesses of the approach you are representing, and decide how you will react to such criticisms.

ANSWER KEY

Multiple Choice

Correct answers are in bold.
Comments related to other options indicate why that response is incorrect.

1. **c)**
    b) It only tests children who are enrolled in school, so it may misrepresent the population.

2. **d)**

3. **d)**

4. **c)**

5. **c)**

6. **d)**

7. **d)**

8. **d)**

9. **b)**

10. **d)**

11. **b)**
    a) Its reliability is not yet known.

12. **a)**

13. **d)**

14. **c)**

15. **a)**

16. **d)**

    b)   Actually, people with higher IQ exerted less mental effort.

17. **c)**

    b)   They often include academic tasks.

18. **c)**

19. **b)**

    d)   Remember, a test cannot be valid unless it is reliable.

20. **c)**

Matching

  _j_ 1. Authentic assessment
  _i_ 2. Choice reaction time
  _a_ 3. Function
  _h_ 4. MRI
  _g_ 5. Neural efficiency
  _c_ 6. PET scan
  _b_ 7. Portfolio
  _e_ 8. Structure

Correct terms for responses left over:
d)   dynamic assessment
f)   NAEP

CHAPTER 15

Teacher Designed Assessments: Traditional and Alternative

LEARNING OBJECTIVES

1. Compare and contrast purposes of classroom assessment.

2. Explain the role of instructional objectives, and identify objectives at different levels of learning.

3. Compare and contrast classroom assessment techniques.

4. Discuss general guidelines for scoring essay tests, and describe two essay-grading approaches.

5. Describe approaches for evaluating multiple-choice items.

6. Describe the sequential stages of testing.

7. Explain how to summarize a class's test performance.

8. Describe good test-taking skills and the implications of their development.

9. Discuss the use of portfolios for assessment in the classroom.

10. Discuss the use of performance assessment in the classroom.

11. Describe guidelines and considerations for grading.

STRENGTHENING WHAT YOU KNOW

The purpose of this chapter is to examine the roles of assessment in the learning process and to evaluate several approaches to classroom assessment. The chapter provides guidelines teachers should consider in developing assessments and grading students.

Objective 1. **Compare and contrast purposes of classroom assessment.**

    A. *Purposes of Assessment*

1. Fill in the matrix with an example of what each type of test would be like (e.g., describe a criterion-referenced, formative test).

		BASIS OF COMPARISON	
		Criterion-referenced	Norm-referenced
INSTRUCTIONAL USE OF TEST INFORMATION	Formative		
	Summative		

2. In the table below, contrast the motivational impact of norm-reference and criterion-referenced assessment.

MOTIVATIONAL IMPACT

Criterion-Referenced Tests	Norm-Referenced Tests

Objective 2. **Explain the role of instructional objectives, and identify objectives at different levels of learning.**

A. *Role of Objectives in Teaching*

Explain why teachers use instructional objectives.

B. *Bloom's Taxonomy of Learning Objectives*

In the graphic, write each level and give a brief definition and original examples.

Objective 3. **Compare and contrast classroom assessment techniques.**

    A.   *Summarizing / Comparing Key Features*

1. Complete the table by filling in key features of each instructional approach.

	Guidelines for Designing	Strengths	Weaknesses
Essay			
Multiple choice			
Matching			
True/False			
Completion/ Fill-in- the-blank			
Portfolios			
Performance Assessment			

Objective 4.   **Discuss general guidelines for scoring essay tests, and describe two essay-grading approaches.**

    A.   *Guidelines for Grading Essays*

1. Explain several ways to increase reliability when grading essay exams.

2. Explain other decisions teachers need to make when grading essays.

    B.   *Analytical Grading of Essays*

1. Describe the analytical grading method.

2. When is this approach most appropriate? Why?

    C.   *Holistic Grading of Essays*

1. Describe the holistic grading method.

2. When is this approach most appropriate? Why?

Objective 5. Describe approaches for evaluating multiple-choice items.

   A.   *Item Difficulty*

1. Explain how item difficulty is calculated.

2. Explain what an item difficulty of .75 means.

   B.   *Item Discrimination*

1. Explain how item discrimination is calculated.

2. Explain what an item discrimination of .25 means.

3. What steps should a teacher take if an item has a discrimination of -.30?

   C.   *Item Distractor Analysis*

1. Explain the purpose of an item distractor analysis.

2. Name two ways to conduct an item distractor analysis.

3. Explain a testing practice that can reduce arguments over unfair distractors.

Objective 6.   **Describe the sequential stages of testing.**

  A.   *Stages In Test Preparation and Administration*

Give a description of how to carry out each of the stages of testing.

Determine purpose

↓

Select/prepare test items

↓

Administer test

↓

Score tests

↓

Give feedback

↓

Revise test

Objective 7.   **Explain how to summarize a class's test performance.**
    A.   *Frequency Distributions*

1. Explain how to create a frequency distribution.

2. Define histogram.

3. Compare the following types of distributions, explaining what each tells about the class composition.

Normal	Positively Skewed	Negatively Skewed	Bi-modal

Objective 8.   **Describe good test-taking skills and the implications of their development.**

A.   *Good Test-Taking Skills*

1. Fill in the chart with descriptions of the general strategies that are useful to test-takers.

Academic Preparation Strategies	Physical Preparation Strategies	Attitude-Improving Strategies	Anxiety-Reducing Strategies	Motivational Strategies

2. List and briefly explain four other specific strategies good test-takers use:

a)

b)

c)

d)

3. What kinds of things can teachers do to help students become effective test-takers?

B.   *Development of Test-Taking Skills*

1. In the chart below, fill in guidelines for test design, based on developmental differences in test-taking.

AGE RANGE	TEST DESIGN IMPLICATIONS
Primary Grades	
Middle Grades	
Upper Elementary	
Junior/Senior High School	
Older Adults	

Objective 9.　　　　**Discuss the use of portfolios for assessment in the classroom.**

　　A.　　*Constructing Portfolios*

1. List three essential steps in constructing an assessment portfolios, and explain why each is important.

a)

b)

c)

2. Based on the content-area portfolio examples, list several features/ideas that you would want to include if you were a teacher using portfolio assessment.

　　B.　　*Evaluating Portfolios*

1. Explain how teachers can use a review sheet to assess student progress over time.

2. Write a brief narrative describing a hypothetical portfolio conference.

C. *Developmental Considerations in Portfolio Assessment*

1. Based on the developmental concerns for young children, write some suggestions for elementary teachers using portfolio assessment in the classroom.

2. Based on the developmental concerns for high school students, write some suggestions for high school teachers using portfolio assessment in the classroom.

Objective 10.   Discuss the use of performance assessment in the classroom.

    A.   *Examples of Performance Assessment*

List as many content areas as you can think of in the left column. In the right column, brainstorm types of performance assessment that could be used in that content area.

Content Area	Examples of Performance Assessment

    B.   *Evaluating Performance*

Compare and contrast the use of checklists and rating scales by describing how each can be used for performance assessment.

Checklists	Rating Scales/Scoring Rubric

345

Objective 11.   **Describe guidelines and considerations for grading.**
   A.   *Principles of Grading*
Illustrate each principle of grading by creating a more hypothetical situation in which the principle is violated. Then suggest how the problem could be avoided or corrected.

Grading Principle	Example of Violation	Suggestions for Avoiding/Correcting Problems
Fairness		
Accuracy		
Consistency		
Defensibility		

   B.   *Standards of Comparison*
1. Explain the instructional implications and potential difficulties with each of the following standards of comparison in grading.

Comparison Standard	Implications	Potential Difficulties
Norm-referenced		
Criterion-referenced		
Percentage cutoffs		
Effort / Improvement		

2. According to the chapter, what factors should **not** be major components of grades?

PRACTICE TESTS

[See answer key at the end of the chapter for correct responses.]

Multiple Choice

Circle the letter of the best response to each question.

1. Which kind of evaluation is typically used for assigning grades?
   a) criterion-referenced
   b) formative
   c) norm-referenced
   d) summative

2. "Grading on a curve" is an example of which type of assessment?
   a) criterion-referenced
   b) formative
   c) norm-referenced
   d) summative

3. For what kind of test is a teacher likely to let students grade their own papers?
   a) criterion-referenced
   b) formative
   c) norm-referenced
   d) summative

4. Your child brings home a report card based on an A, B, C, D, F letter-grade system. Your child has a "D" in one subject area, and the teacher has enclosed a note expressing concern that your child is not doing well according to his criterion-referenced evaluations. What does this mean about your child's performance?
   a) Her performance was a little bit worse than her classmates'.
   b) Her performance did not meet the teacher's acceptable requirements.
   c) The teacher was probably grading to identify a select group of students who require assistance in his class.
   d) This is not her final grade, just a way of giving feedback about how she is doing so far so that she can make changes and do better.

5. Which of the following is an appropriate conclusion/decision for a norm-referenced test?
   a) This examinee did not meet the cut-off on the written test, so he will not receive his driver's license.
   b) Because the student got an average score, he was not admitted to the gifted and talented program, which only has a few openings.
   c) The student got the highest possible grade, so she clearly got most of the items correct.
   d) The fact that this person got the lowest possible grade obviously shows she did not understand the information.

6. Which feature of testing is most likely to promote life-long learning?
   a) criterion-referenced
   b) formative
   c) norm-referenced
   d) summative

7. In an educational psychology syllabus, one of the objectives says, "Students will design a lesson that combines several affective approaches to teaching mathematics." What cognitive level does this objective represent?
   a) Application
   b) Analysis
   c) Evaluation
   d) Synthesis

8. An educational psychology test item says, "Write the names of the six levels of Bloom's taxononmy." What cognitive level does this item represent?
   a) Application
   b) Analysis
   c) Comprehension
   d) Knowledge

9. An educational psychology student is given a transcript of a reading lesson and is asked to identify aspects of the lesson that are consistent or inconsistent with transactional strategies instruction. What level of learning does this task represent?
   a) Application
   b) Analysis
   c) Comprehension
   d) Evaluation

10. Which of the following is a good practice for constructing essay items?
    a) Let students choose from a few possible questions.
    b) Indicate how long students should spend writing their response.
    c) Use open-ended questions, such as one that begins, "Discuss your understanding of..."
    d) Give only one or two questions so that students can answer in depth.

11. Which of the following is an effective practice for grading essay tests?
    a) Divide the papers into stacks for high, average, and low performance after a brief review.
    b) Read one person's exam all the way through before going to the next exam.
    c) Have students put their names at the top of each test page so the papers will not get mixed up or lost.
    d) Take frequent breaks and spread the grading over a few sessions to avoid fatigue.

12. Which of the following best describes the analytic method of grading essay tests?
    a) Students know what the general grading criteria will be before writing their responses.
    b) Mistakes are weighed more heavily than missing ideas.
    c) Students receive a two-part grade--a "content" grade and a "form" grade.
    d) Students are given points for including each idea that is in a teacher-constructed essay.

13. Which of the following is an effective practice in constructing multiple choice tests?
    a) Keep the number of distractors as small as possible.
    b) Use trick questions to prevent guessing and insure that students read the whole item.
    c) Use the option "none of the above" as an alternative.
    d) Make each question independent of the other questions.

14. Which of the following is true of multiple-choice tests?
    a) They are limited to the lower levels on Bloom's taxonomy.
    b) Scoring is often unreliable.
    c) They require continuous revision.
    d) They are easy to construct.

15. A teacher finds that the item difficulty on an item is .25; which of the following does the teacher know about the item?
    a) 25% of the students missed the item.
    b) This is a good item for a norm-referenced test.
    c) This was a difficult item on the test.
    d) Students who scored high on the test probably got the item right, and students who scored low on the test probably got it wrong.

16. Students are arguing that a test was too hard because there were a lot of tricky items. During a discussion of the test, students argue that their choices are plausible. What kind of item analysis would best help a teacher resolve such arguments?
    a) Item difficulty
    b) Item discrimination
    c) Item distractor
    d) Item reliability

17. Which kind of testing is limited to testing a small percentage of content?
    a) Essays
    b) Matching
    c) Multiple choice
    d) True-false

18. Which of the following is a good practice for matching tests?
    a) Put the responses in random order.
    b) Include 20 or 30 items.
    c) Make sure responses can be used more than once.
    d) Include more response choices than premises/items.

19. Which of the following is a good practice for true-false tests?
    a) State some items negatively so students have to think about them.
    b) Have the same number of true and false statements.
    c) Include more than one idea in the question, so that the student has to decide if both parts are true.
    d) Use absolute terms like "every."

20. Which of the following teacher behaviors most directly supports "academic preparation strategies" for test-taking?
    a) Teaching students how to ask the right questions to get valuable information about the test content and structure.
    b) Telling students the test date well in advance so they aren't as likely to "cram."
    c) Telling students that the test is easy and everyone will probably get an A.
    d) Teaching students to calm themselves down when they "draw a blank" during a test.

21. Which of the following test practices addresses the test skills of second-grade students?
    a) Use several item formats on the test to give students different ways of showing what they know.
    b) Give students unlimited time to finish their tests because they need extra time to think about the items.
    c) In multiple-choice items, give four or five choices to prevent guessing.
    d) Give students tests that you expect to take them half an hour or less.

22. You are a teacher trying out portfolio assessment with your students. Which of the following guidelines would be most important to give students?
    a) The portfolio should be a collection of your most recent work.
    b) Limit the portfolio to your best work in this class.
    c) Here is a list of the items you will include in your portfolio....
    d) Write a note about how each piece represents your learning.

23. Mrs. Evanel believes that improving students' learning processes is even more important than the products of their learning. What testing approach would best show Mrs. Evanel how her students are processing information?
    a) Essay tests, graded holistically
    b) Essay tests, graded analytically
    c) Performance assessment
    d) Portfolios

24. According to the chapter, which of the following is most defensible to consider as a major component in grading a student?
    a) Formative assignments
    b) How well other students did
    c) Effort
    d) Participation

Completion

Fill in each blank with the best fitting term from the chapter. Terms are used only once.

1.   Ego-involved, performance orientations toward learning are most likely to result from _____ evaluations.

2.   Tests considered less open to interpretation in grading are called _____.

3.   One way to represent how a test will proportionately cover content and cognitive performance is in a _____.

4.   Test scores across a group can be summarized graphically in a _____.

5.   A list of how many people earned each score is called a _____.

6.   Although educational literature disusses its use, there is little research evaluating _____ assessment.

7.   Students talk with teachers about the strengths and weaknesses of a collection of their school work in a _____.

8.   Students are required to directly demonstrate their knowledge and skills in _____ assessments.

Matching

Match the letters of the description on the right with the corresponding numbered terms
on the left. Use each description only once. Some descriptions may be left over.

_____	1.	Criterion-referenced
_____	2.	Formative
_____	3.	Instructional objective
_____	4.	Item difficulty
_____	5.	Item discrimination
_____	6.	Negatively skewed
_____	7.	Normal distribution
_____	8.	Norm-referenced
_____	9.	Positively skewed
_____	10.	Scoring rubric
_____	11.	Summative

a) Percentage of students who missed the item
b) Compared to other students
c) Percentage of students who answered correctly
d) A likely distribution for a class that has two distinct ability groups
e) Formal evaluation of student learning
f) How well a test question can tell high test scorers from low test scorers
g) Compared to an acceptable standard of what should be learned
h) Distribution of a difficult test
i) Desired change in a student
j) Feedback to students about what they have learned and to teachers about what students need to know
k) Distribution of an easy test
l) Bell-shaped curve
m) Continuum that can be used for used in performance assessment

LEARNING STRATEGIES

Below are examples of strategies that can help you understand major chapter concepts. Use these examples to guide your own strategies.

Strategy Example #1
Create your own instructional objectives for your learning in this course.
Determine the level of each objective on Bloom's taxonomy.
Add objectives so that all levels are included.
Design ways to self-test whether you have met each of your objectives.

Strategy Example #2
To remember the difference between positively and negatively skewed distributions, visualize a number line, which has positive or larger numbers to the right and negative or smaller numbers to the left. The critical point to remember is that "positive" or "negative" refers to the direction the "tail" is pointing (where the scores taper off), rather than where the scores are most clustered. A positively skewed distrubution trails off as the scores get higher, meaning that most of the scores were low.

Strategy Example #3
Personalize the principles of grading by recalling any instances when you argued about a grade or questioned how a paper or course was graded. Which of the principles represent your greatest concerns? [Fairness, Accuracy, Consistency, Defensibility]

Strategy Example #4
Rate your study habits on each of the test-taking strategies described in the chapter.
Prepare for your next exam by developing a plan based on all the strategies.

ANSWER KEY

Multiple Choice

Correct answers are in bold.
Comments related to other options indicate why that response is incorrect.

1. **d)** could be either criterion-referenced or norm-referenced

2. **c)**

3. **b)**

4. **b)**
   a) norm-referenced
   c) norm-referenced
   d) formative

5. **b)** Norm-referenced tests are often used to predict future success and to select a few students from a large group.
   a) If the cutoff is norm-referenced only a certain percentage of students could pass. If everyone does very well on the test, this approach could deny some very knowledgable drivers a license.
   d) See a).

6. **a)**

7. **d)** beyond application because it requires the learner to integrate several approaches

8. **d)**

9. **b)**

10. **b)**

11. **a)**

12. **d)**
   a) A good practice that could apply to either analytical or holistic approaches.
   b) Not necessarily; in fact, the reverse is more likely to be a problem.
   c) This could also be true of holistic grading.

13. **d)**

14. **c)**
   b) Reliability is an advantage of multiple-choice over essay tests.

15. **c)**
   d) item discrimination

16. **c)**

17. **a)**

18. **d)**
   c) Not necessarily (no particular advantage), but do make sure to tell students if this is the case.

19. **b)**

20. **a)**
   b) targetted physical preparation
   c) poor attempt at reducing anxiety--probably reduces effort, but not anxiety
   d) anxiety-reducing strategy

21. **d)**
   a) Students may spend too much mental capacity trying to adjust to each question format.
   b) Younger students need shorter tests.
   c) Young students require fewer options so they don't become overwhelmed. Remember their working-memory capacity is limited.

22. **d)**

23. **c)**

24. **b)** Norm-referenced tests, though arguable, are still considered defensible.
   a) Students could be penalized for early misunderstandings, even though they later master the content.

Completion

1. norm-referenced
2. objective
3. table of specifications
4. histogram
5. frequency distribution
6. portfolio
7. portfolio conference
8. performance assessments

Matching

_g_ 1. Criterion-referenced
_j_ 2. Formative
_i_ 3. Instructional objective
_c_ 4. Item difficulty
_f_ 5. Item discrimination
_k_ 6. Negatively skewed
_l_ 7. Normal distribution
_b_ 8. Norm-referenced
_h_ 9. Positively skewed
_m_ 10. Scoring rubric
_e_ 11. Summative

Correct terms for responses left over:
a) [none--distractor for item difficulty]
d) bi-modal

CHAPTER 16

Learner Diversity

LEARNING OBJECTIVES

1. Compare three classifications of exceptional learners, focusing on information processing differences and instructional needs.

2. Explore instruction designed to serve students with disabilities.

3. Discuss Down Syndrome as a common severe mental disability.

4. Explain characteristics and treatment of Attention Deficit Hyperactivity Disorder.

5. Compare two types of giftedness.

6. Explore environmental factors that can place students at risk for school difficulties.

7. Explore biological factors that can place students at risk for school difficulties.

8. Discuss gender differences in cognition.

9. Trace changes in thinking abilities across the lifespan.

STRENGTHENING WHAT YOU KNOW

The purpose of this chapter is to introduce several factors that underlie individual differences in learning. The chapter discusses the learning needs of students based on these individual differences, suggesting effective instructional approaches.

Objective 1.   **Compare three classifications of exceptional learners, focusing on information processing differences and instructional needs.**

	Mental Retardation	Learning Disability	Giftedness
Definition(s) / Characteristics			
% of People			
Contributing or Supporting Factors / Causes			
Information Processing			
Instructional Approaches/ Implications			

Objective 2.   **Explore instruction designed to serve students with disabilities.**

   A.   *Past and Present Ways Schools Have Approached Disabilities*

1. Compare and contrast past and present approaches to serving students with disbilities.

Past Approaches	Present Approaches

   B.   *IEP*

1. In the concept map below, identify the components of an Individualized Education Program for a student with disabilities.

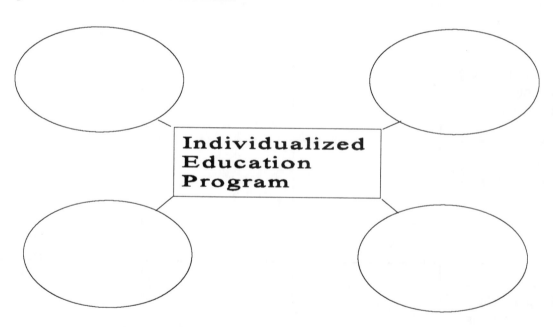

Objective 3.   **Discuss Down Syndrome as a common severe mental disability.**
    A.   *Symptoms*
1. Describe the symptoms of Down Syndrome.

    B.   *Instructional Implications*
1. What are the main instructional implications for students with Down Syndrome?

Objective 4.   **Explain characteristics and treatment of Attention Deficit**
                             Hyperactivity Disorder.
    A.   *Symptoms of ADHD*
1. Describe three main categories of symptoms of ADHD
        a)

        b)

        c)

2. Explain two key features of accurate diagnosis of ADHD.
        a)

        b)

    B.   *Treating Students Who Have ADHD*
1. Describe strengths and weaknesses of using medications to treat ADHD.

2. Describe instructional approaches that can help students who have ADHD.

Objective 5.   **Compare two types of giftedness.**

    A.   *Types of Giftedness*

1. Compare the following two types of giftedness by describing their characteristics:

Prodigies	People with High General Intelligence

2. Summarize environmental factors that support prodigious genius.

Objective 6.   **Explore environmental factors that can place students at risk for school difficulties.**

    A.   *Poverty*

1. List key problems poor children face in school learning.

2. Contrast traditional and current approaches to teaching disadvantaged students.

Traditional instructional approach (Explain why it is ineffective.)	Current instructional approach

362

B.   *Culture and Disadvantaged Minorities*

1. Explain potential "mismatch" problems between minority cultures and school culture.

2. Summarize recommendations for serving disadvantaged minorities in schools.

C.   *Non-Native Speakers of the Majority Language*

1. What challenges do non-native speakers face in school?

2. Compare and contrast three types of instructional programs for speakers of minority languages, describing the emphasis of each type of program:

Bilingual Education	English as a Second Language (ESL or ESOL)	Bilingual Immersion (also called Dual Immersion)

Objective 7. **Explore biological factors that can place students at risk for school difficulties.**

   A.   *Comparing Impacts and Interventions*

1. In the table below, compare several diseases/chemical assaults that can affect learning.

	Impacts on Learning	Interventions (especially for learning)
AIDS/HIV (Human Immunodeficiency Virus)		
Cancer (list impacts in order from minimum to maximum)		
FAS (Fetal Alchohol Syndrome)		
Cocaine/ other illegal drug exposure		
Lead exposure		

Objective 8. **Discuss gender differences in cognition.**

    A. *Intellectual Processing and Abilities for Males and Females*

1. In the table below, define each characteristic and compare males and females on each.

	Females	Males
Intelligence (IQ)		
Verbal ability		
Visual-spatial skills		
Field independence/ dependence		
Quantitative skills		
Strategies / Metacognition/ Knowledge		
Motivation		

2. Overall, how great are the intellectual differences between females and males?

B. *Theories of Processing Differences Between Males and Females*

1. Summarize four biological theories of gender differences in intellectual skills.

a)

b)

c)

d)

2. List key points of psychosocial theories of gender differences in intellectual skills.

Objective 9. **Trace changes in thinking abilities across the lifespan.**

1. Why is it important to understand the learning abilities of older adults?

2. Explain the findings of studies of memory strategies with older adults.

3. Describe the results of older adult questionnaires on metamemory.

4. Explain the possible roles of motivation in memory differences of older adults.

PRACTICE TESTS

[See answer key at the end of the chapter for correct responses.]

Multiple Choice

Circle the letter of the best response to each question.

1. Which of the following instructional approaches does current legislation require for students with learning disabilities?
   a) Self-contained classrooms
   b) Inclusion
   c) Resource room instruction
   d) Individualized education program

2. Which of the following is a feature of mild retardation?
   a) IQ is one or more standard deviations below average.
   b) Difficulties may not be apparent until adulthood.
   c) Dependence on family or social services for most of one's life.
   d) Difficulties with social skills.

3. Which of the following is true of learning disabilities?
   a) Dislexia is the most common learning disability.
   b) More than 5% of the student population have learning disabilities.
   c) Students with learning disabilities have below average IQs.
   d) A student with learning disabilities is average or above in some subjects.

4. Which of the following is true of ADHD (Attention Deficit Hyperactivity Disorder)?
   a) A student who is hyperactive at school often behaves normally at home.
   b) Attention deficits are easier to diagnose when the student is in the upper elementary grades.
   c) Ritalin and other medications for ADHD are depressants that keep children calm so they can focus.
   d) Medications may interfere with students' learning to control their behavior and learning.

5. For which of the following student populations is it considered effective and appropriate to offer a reinforcer, such as money, for accepted behavior?
   a) Gifted
   b) Mentally retarded
   c) Learning disabled
   d) Living in poverty

6. Which of the following is true of Down Syndrome?
   a) It is a rare form of severe mental disability.
   b) It is impossible to tell by looking which child in a classroom has Down Syndrome.
   c) Children with Down Syndrome function best in institutions where they can receive appropriate care for their disability.
   d) Adults are less interactive with children who have Down Syndrome, expecting the child to need lots of assistance.

7. Which of the following is true of giftedness?
   a) The definition of giftedness is an IQ of 130 or higher.
   b) Geniuses are often socially awkward.
   c) Geniuses tend to put more importance on their career than on their personal lives.
   d) Environment plays a major role in developing genius.

8. Which of the following is best supported as a way to help a child prodigy develop her genius?
   a) Guide her toward a well-established field.
   b) Encourage her to be well-rounded, rather than focusing attention on one specific area.
   c) Find a mentor in the community who has expertise in her area of interest.
   d) Avoid drawing attention to her talent, to prevent embarrassment.

9. Which of the following is true of students who are at risk of school failure?
   a) They tend to be submissive to authority figures.
   b) There is little that teachers or schools can do to help them.
   c) Among students with a particular risk (e.g., poverty), there is little variation in functioning.
   d) Interventions are best when introduced as early as possible.

10. Which of the following best represents current recommendations for economically disadvantaged students who are having difficulties in school?
    a) Work on higher-order thinking.
    b) Use guided discovery to make learning more motivating.
    c) Focus instruction on the basic skills these students are lacking.
    d) Use mastery learning, breaking tasks down into units students can handle.

11. Which of the following is recommended in helping disadvantaged minorities?
    a) Encourage parents to teach students how to behave in school.
    b) Grade students based on improvement.
    c) Encourage students to use standard English in the classroom.
    d) Work on remediating students' weaknesses.

12. Which of the following is true of students whose native language is not the majority language?
    a) Parents encourage students to learn the majority language.
    b) Legal pressures are forcing schools to teach in the majority language (e.g., English).
    c) The students often lack basic skills, like students living in poverty.
    d) The goal of instruction is to minimize use of the native language.

13. Which of the following best describes the emphasis of bilingual immersion (dual immersion) programs?
    a) Teaching students in their native language, then reinforcing with some use of the target language (e.g., English).
    b) Encouraging use of the target language (e.g., English) in the classroom, but encouraging students to speak their native language at home.
    c) Fostering proficiency in both the native language and the target language.
    d) Teaching students entirely in their second language.

14. Children with which of the following conditions experience problems ranging from missed school to fatigue to brain impairment?
    a) AIDS/HIV
    b) Cancer
    c) Cocaine/polydrug exposure
    d) Fetal Alcohol Syndrome
    e) Lead Exposure

15. Cognitive stimulation can be effective with students having which condition?
    a)   AIDS/HIV
    b)   Cancer
    c)   Cocaine/polydrug exposure
    d)   Fetal Alcohol Syndrome
    e)   Lead Exposure

16. Which condition would most likely lead students to lie and steal?
    a)   AIDS/HIV
    b)   Cancer
    c)   Cocaine/polydrug exposure
    d)   Fetal Alcohol Syndrome
    e)   Lead Exposure

17. With which condition is it critical to avoid too much stimulation?
    a)   AIDS/HIV
    b)   Cancer
    c)   Cocaine/polydrug exposure
    d)   Fetal Alcohol Syndrome
    e)   Lead Exposure

18. Which intellectual capacities starts stronger in females, but becomes stronger in males by adolescence?
    a)   IQ
    b)   Verbal ability
    c)   Visual-spatial skills
    d)   Quantitative skills

19. Which of the following does **not** represent a biological theory of gender differences in cognition presented in the chapter?
    a)   Males have stronger connections between the two hemispheres of the brain than females do.
    b)   In females, extra verbal capacity "squeezes out space" typically dedicated to spatial skills.
    c)   A woman having her menstrual period performs better on typically male-dominated tasks.
    d)   Testosterone affects brain development.

20. Which of the following represents a psychosocial theory of gender differences in cognition?
    a)   Math teachers expect more from girls than boys.
    b)   Stereotypes give more value to female roles than male roles.
    c)   Aggression is more supported in science classrooms than cooperation.
    d)   Adults trust girls more than boys, giving girls more freedom to explore.

21. Which of the following statements accurately represents a characteristic of older adults' thought processing?
    a)   They do not benefit as much from strategies instruction.
    b)   They stay more focused on the task than younger adults.
    c)   They are overly optimistic about their mental capabilities.
    d)   They are less likely to use their knowledge to organize information.

Completion

Fill in each blank with the best fitting term from the chapter. Terms are used only once.

1. Public Law 94-142 requires that students with disabilities are educated in the _____.

2. A plan for the student with special needs, including summary of performance, goals, support services, and evaluation approaches, is called the _____.

3. It is difficult to distinguish social-cultural retardation from organic factors because social-cultural retardation may be _____.

4. A student who is especially fidgety, makes careless mistakes, behaves inappropriately, and often doesn't finish homework may have _____.

5. A person who has extreme talent in a specific area is referred to as a _____.

6. Gifted students' thinking parallels the model of _____.

7. A chemical assault that could be prevented through home repair is _____.

8. Withdrawing when challenged or going out of control when overwhelmed are possible symptoms of _____.

9. Females are more likely than males to be _____, perceiving patterns as wholes rather than separate items.

Matching

Match the letters of the description on the right with the corresponding numbered terms on the left. Use each description only once. Some descriptions may be left over.

_____ 1. Dyslexia
_____ 2. ESL
_____ 3. Field independence
_____ 4. Inclusion
_____ 5. Impulsive
_____ 6. Resource room
_____ 7. Self-contained

a) Encourages students to learn in their native language
b) Ability to analytically perceive an object apart from its background
c) Separate classroom taught by special education teachers
d) Viewing patterns holistically
e) Students with disabilites receive all instruction in a regular classroom
f) Emphasizes using the majority language while learning content
g) Responds quickly without thinking through alternatives
h) Inability to decode words, despite instruction
i) Temporary removal from the classroom for special instruction

LEARNING STRATEGIES

Below are examples of strategies that can help you understand major chapter concepts. Use these examples to guide your own strategies.

Strategy Example #1
Choose one or more of the individual differences (one with which you are less familiar). Role-play being a student from that population, focusing on how you might behave in class or a particular difficulty you might have.

Strategy Example #2
Working with a partner or a study group, assign each person one of the theories of male/female processing differences. Hold a debate between biological/psychosocial theories, or among particular viewpoints.

Strategy Example #3
Sketch out a lesson plan for a diverse classroom that includes at least one student from each of the groups described in the chapter (not entirely unlikely in many urban areas). Note how aspects of the lesson are designed to meet the needs of particular students.

ANSWER KEY

Multiple Choice

Correct answers are in bold.
Comments related to other options indicate why that response is incorrect.

1. **d)**

2. **d)**
   a) two or more

3. **d)**

4. **d)**
   b) ADHD symptoms should appear by age 7.
   c) These are stimulants that keep areas of the brain active to control attention.

5. **b)**

6. **d)** This undermines children's risk-taking and leads them to prefer easy tasks.
   a) Most common
   c) Students with Down Syndrome often function best in supportive home environment.

7. **d)**
   a) IQ is only one of the criteria, and different cutoff points are sometimes used.

8. **c)**

9. **d)**
   a) They tend to be noncompliant.

10. **a)**

11. **b)**
    c) Be sensitive to possible misunderstandings due to dialect and language.
    d) Focus on building on strengths.

12. **c)**
   a) Some parents may have negative attitudes about the majority language.
   b) Legal emphasis is on increasing bilingual education.

13. **c)**

14. **b)**

15. **a)**

16. **d)**

17. **c)**

18. **d)**

19. **a)** The reverse may be true.

20. **c)**

21. **d)**

Completion

1. least restrictive environment
2. Individualized Education Program
3. hereditary
4. ADHD--attention deficict hyperactivity disorder
5. prodigy
6. the good information processor
7. lead exposure
8. prenatal crack-cocaine exposure
9. field-dependent

Matching

__h__	1.	Dyslexia
__f__	2.	ESL
__b__	3.	Field independence
__e__	4.	Inclusion
__g__	5.	Impulsive
__i__	6.	Resource room
__c__	7.	Self-contained

Correct terms for responses left over:

a) bilingual

d) field dependence